World weather and climate

World weather and climate

DENIS RILEY
Senior Tutor
Thornbridge School, Sheffield

LEWIS SPOLTON
Senior Lecturer in Education
University College, Swansea

CAMBRIDGE
UNIVERSITY PRESS

Published by the Syndics of the
Cambridge University Press
Bentley House, 200 Euston Road, London NW1 2DB
American Branch: 32 East 57th Street, New York, N.Y.10022

Library of Congress Catalog Card Number: 73–75858

ISBNs:
0 521 21076 4 hard cover
0 521 21075 paperback

First published 1974

Printed in Great Britain
at the University Printing House, Cambridge
(Brooke Crutchley, University Printer)

Contents

Acknowledgements

Thanks are due to the following people and
organisations for permission to reproduce photographs
and diagrams: Meteorological Office, Edmonton,
Canada, 3, 4, 75, 77, 106; J. M. Walker, 5; H. H. Lamb,
9, 20, 95; R. K. Pilsbury, 15, 23, 25, 28, 32, 108;
The Controller, Her Majesty's Stationery Office, 22, 24,
26, 30, 31, 49, 57 (with D. E. Pedgley), 61, 64, 66, 68,
69, 72, 73, 74, 92, 93, 96, 97 (with D. E. Pedgley), 98
(with D. E. Pedgley), 99, 127; Royal Meteorological
Society, 10, 41, 42, 43, 44, 46, 55(a) and (b), 100,
101, 109; Ambassador College, 39; Professor
S. Gregory, 51, 111; Indian Meteorological Service, 52,
53, 54; New Zealand Meteorological Office, 78, 79;
U.S. Department of Commerce Monthly Weather
Review, 47(a) and (b), 48, 59, 71, 80, 90, 91, 118,
133(a) and (b); Meteorological Office, U.S.S.R. 81;
Japanese Meteorological Agency, 82, 83; U.S. Weather
Bureau, 85, 86, 87; New Orleans Weather Bureau, 89;
Department of Transport, Republic of South Africa, 94;
P. A. Records, 103; Swiss Meteorological Institute, 104,
105; The Curator, Weston Park Museum, Sheffield,
data for 62, 63, 65, 67.

The data for the climatic graphs are taken mainly from
Her Majesty's Stationery Office Tables of Temperature,
Relative Humidity and Precipitation for the World.

The cover picture on the paperback is a satellite
infra-red photograph of clouds in hurricane 'Debbie'
(19 August 1969) which clearly shows the eye of the
storm. *Photo:* NASA.

Preface

Meteorology on a world scale combines a study of the physics of the atmosphere with geography. Geography studies the areal distribution of phenomena over the earth's surface and is concerned with similarities and differences between areas. The distribution of atmospheric phenomena over the earth is differentiated by latitude, topography, altitude, the distribution of land, sea and ice-caps, and also, to some extent, by forests and great cities. All these influences are essentially geographical. The physicist approaches the study of atmosphere by considering what happens under ideal conditions and then makes allowances for deviations which occur due to geographical circumstances. The geographer makes more use of the approach from reality. He studies the synoptic chart which depicts actual conditions and uses the insights supplied by the physicist in an attempt to understand what is happening. The approaches are complementary. This book, written for geographers, makes use of the general basic principles of physics. They are explained in a way suitable for geographers in Part I which consists of the first five chapters.

Part II builds on the principles established in the first part and provides a description and explanation of regional weather types. An important feature is the provision of synoptic charts which illustrate the weather at a particular time. Typical synoptic charts for many areas of the world are included. Each of the continents is represented. The text cannot always draw attention to every detail of the synoptic situations. Users of the book can do much self-help by comparing conditions between charts. Many exercises can be devised. Tracing paper placed over the maps allows isotherms to be drawn, areas of cloudiness or precipitation to be shaded in, and the resulting patterns can be discussed and explained. Because weather and climate are three-dimensional studies, upper-air data is provided when it is useful. In studying the charts an attempt should be made to visualise the conditions in depth. This is not easy but use of the many diagrams showing conditions in the upper air will improve this facility.

Part III is devoted to climatology. However, if the subject is thought of merely in terms of averages much vitality is lost. A study of climate needs to take account of extreme conditions and variable conditions. The climatic graphs are drawn to include a wealth of information and are discussed in dynamic terms using the ideas established in Parts I and II.

Geography is sometimes thought to be concerned only with environment, in our case, the atmosphere, as it concerns man. Although most of what is discussed and explained in this book concerns man and affects his mode of living, geography is not necessarily restricted to this. A useful distinction has been made between the phenomenal environment which includes all natural phenomena and the works of man, and the behavioural environment which includes only those aspects of the phenomenal environment which affect man. The upper layers of the atmosphere have always been part of the phenomenal environment. With the coming of radio these upper layers are useful since they reflect radio waves and enable transmissions to be made over long distances. Thus they become part of the behavioural environment. Man is most affected by the weather in the lowest layers, and the book concentrates on the behavioural environment, though when necessary it will describe events which are still only phenomenal.

The intention is not merely to provide a text which can be read and absorbed but to provide actual evidence which can be analysed and discussed. Only in this way can world patterns of weather and climate really be appreciated.

DENIS RILEY
LEWIS SPOLTON

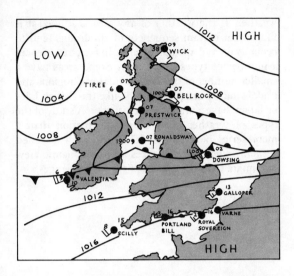

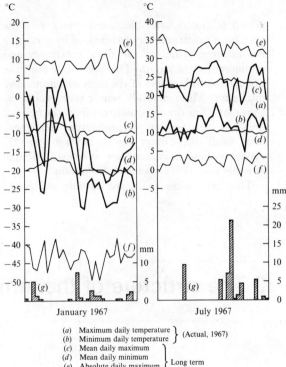

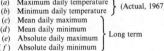

January 1967   July 1967

(a)	Maximum daily temperature	} (Actual, 1967)
(b)	Minimum daily temperature	
(c)	Mean daily maximum	
(d)	Mean daily minimum	
(e)	Absolute daily maximum	} Long term
(f)	Absolute daily minimum	
(g)	Rainfall (Actual, 1967)	

17 July had the highest minimum (17°C). Also shown on the graph are the mean maximum and the mean minimum for each day. To get these figures the minimum for each day is averaged over a number of years and the maximum is treated similarly. The graph also shows the highest and lowest values ever recorded on each specific day during the period of the observations. It can be seen that July 1967 did not break any records though the minimum on 25 July was within 1°C of the extreme minimum recorded for that day. If all the figures were averaged a monthly mean temperature of 16°C would be realised.

The January graph shows similar information for a winter month. Even greater variability is evident. This reflects the greater contrasts in air mass which affect the prairies in winter. Cold air streams from the north can be replaced by Chinook air warmed by descent in the lee of the Rockies or warmer air from a southerly point. The monthly mean temperature is about −16°C. January and July are equally poised above and below zero. However the graphs show that temperatures above freezing point are not uncommon but neither are temperatures below −40°C.

The bar graphs of daily rainfall show eleven days of rain and twenty days without rain in July and sixteen days of rain/snow in January. On the two wettest days in each month the drop in temperatures indicates the passage of cold fronts. Average rainfall figures hide this information yet obviously not every location can be dealt with in the kind of detail given for these months at Edmonton.

However in order to give some idea of possible variations temperature graphs in this book normally

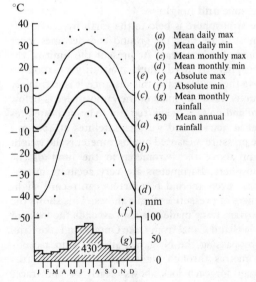

(a)	Mean daily max
(b)	Mean daily min
(c)	Mean monthly max
(d)	Mean monthly min
(e)	Absolute max
(f)	Absolute min
(g)	Mean monthly rainfall
430	Mean annual rainfall

EDMONTON 53° N. 122° W.
Altitude 682 m

give six sets of figures. Edmonton (Fig. 4) is given as an example. The two sets of dots give the highest and lowest temperatures ever recorded in each month: they are the absolute extremes for the entire

3

period of the record. The fainter lines join the monthly maxima and minima obtained by averaging the extreme values of the daily maximum and minimum for the month. The bolder line joins the average monthly values of the daily maximum and minimum. Mean monthly temperature could be shown by a line drawn equidistant from the two full lines. At locations where temperature ranges are very small the fainter lines showing monthly maximum and minimum are omitted. Average monthly rainfall is shown by a histogram.

This book aims to regard climate as typical weather. If the weather of any area is studied it is found that certain sequences of conditions tend to occur again and again. Any area, in fact, experiences a number of types of weather. Some areas have a smaller number of types, e.g. equatorial regions and trade wind deserts, but others, like the British Isles or Japan, have a much greater variety. A study of the types of weather gives more information than a mere series of averages. This approach is called *synoptic climatology*. It takes a more dynamic view of climate than the approach through averages.

2 The structure of the atmosphere

Pressure and height
The atmosphere is held to the earth by gravity. It is most dense near the ground and becomes more tenuous further away. At present the limit of the atmosphere is fixed quite arbitrarily at 1000 km above the earth, and at this distance it is extremely tenuous. Half the mass of the atmosphere is below 5 km and 99% is below 100 km. (Manned satellites orbit at 300 km and weather satellites at 400 km.) The pressure measured by a barometer is the weight of air above the barometer to the limit of the atmosphere. Barometers are very accurate instruments: even aneroid barometers can record slight changes of pressure as Fig. 5 shows. This shows the barogram trace made as a ship ascends the locks of the Welland Canal from Lake Ontario to Lake Erie. In by-passing the Niagara Falls the ship ascends 100 metres through eight locks. The 12 metres average for each lock shows clearly on the trace, indicating that pressures and heights in the atmosphere can be registered in detail.

At sea level pressure varies from about 1080 mbar in an intense anticyclone (see the Siberian anticyclone with a central pressure of 1070 mbar in Fig. 81) to below 900 mbar in a very deep tropical storm (see hurricane Esther with a central pressure of 927 mbar in Fig. 59). Average pressure is about 1013 mbar. Meteorologists often work with constant pressure levels rather than constant heights especially in the upper atmosphere. Some maps in this book show conditions at the 500 mbar level. When variations in the height of this surface are plotted a contour map results. The average height of the 500 mbar surface is 5500 metres. Contours on the surface can be treated as isobars.

Vertical temperature distribution
While pressure decreases steadily with height, temperature shows a much more complicated distribution. In fact as more is discovered about the upper atmosphere the more complicated the distribution is found to be. Snow-covered mountain tops in the tropics suggested to very early observers that temperature decreased with height away from sea level, and until 1890 it was thought the decrease continued right out into outer space. Then balloons carrying thermometers showed that above 10 km temperature ceased to fall but remained steady or even increased. This kink in the temperature curve is clearly shown on the temperature–height diagram in Fig. 42. A bend on the curve indicating a change

5 Barogram from a ship traversing the locks of the Welland Canal, and ascending 100 m from Lake Ontario to Lake Erie

6 Generalised vertical temperature distribution of the atmosphere

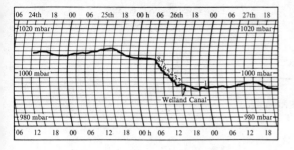

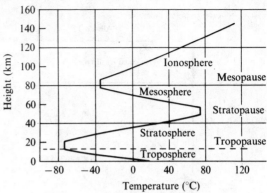

from decrease with height to increase with height is called an *inversion*. Below the inversion is the region called the *troposphere*. Where the inversion occurs is the upper limit of the troposphere, known as the *tropopause*: above is the *stratosphere*. At the equator the troposphere is 16 km deep and at the poles it is about 8 km deep. On average the tropopause shows a gradual slope from equator to poles, but especially in mid-latitudes the height of the tropopause varies with the weather situation. Most of the moisture in the atmosphere is in the troposphere and almost all 'weather' occurs in this lower layer, which is the main concern of this book. As temperature decreases with height in the troposphere and as the troposphere is deepest at the equator, low temperatures in the troposphere occur above the equator, where −80° c may be registered as against −50° c at the polar tropopause. The tropopause used to be thought of as a kind of lid on our weather but the situation is now known to be less simple than that as will be explained later.

In this century and particularly since the Second World War, rockets and, more recently, satellites have shown two cold layers in the atmosphere separated by a warmer band at a height of about 50 km. The names given to the layers of atmosphere and their boundaries are shown on Fig. 6. The outer region of the atmosphere is now known to affect not only the space travellers who pass through it, but also conditions on earth. For instance, oxygen in the upper stratosphere absorbs ultra-violet radiation. In the process ozone is formed and the stratosphere is heated. Since strong ultra-violet radiation is dangerous to human beings, life as we know it on earth may depend on this contingency.

As far as is known at present temperature increases above the mesopause and goes on increasing through the thermosphere with its belts of variable electron density. Eventually the earth's atmosphere

in some extremely attenuated form merges with that of the sun which is the source of the energy on earth.

Radiation and insolation

The temperature of the sun's surface is about 6000° c. Because energy is radiated from a body at a wavelength inversely proportional to its temperature, short-wave radiation is emitted from a high temperature body like the sun. The earth only intercepts a small amount of the sun's radiation. If the energy received from the sun in *one day* is taken as a unit, the energy used on earth by man in a year is $\frac{1}{100}$ of a unit: a large depression uses $\frac{1}{1000}$ of a unit. A hurricane uses $\frac{1}{10}$ as much as a depression; a nuclear bomb about $\frac{1}{10}$ as much as a hurricane. Expressed in another way, a day's energy from the sun is sufficient for 1000 depressions; 10,000 hurricanes or 100 million thunderstorms.

Of the solar energy or heat, which is the way energy is transmitted, reaching the outer edge of the atmosphere, less than half usefully reaches the earth. 19% is absorbed by the gases of the atmosphere: some, as already noted, in the ozone layer; some by water vapour and carbon dioxide; some of it is absorbed on reflection from clouds and impurities. 39% is reflected back into space directly from cloud tops or from the earth's surface. Fresh snow is a very good reflector, reflecting 90%; dense forests reflect 5%; grassland and crops generally about 20%. Reflection from water depends on the angle of the sun. In the morning and evening, at low angles, much more is reflected than at mid-day. Satellite photographs show differential reflection including glitter from the sea. The percentage of radiation reflected is called the *albedo*. Astronauts see the earth as a shining sphere because of its albedo. For ordinary observers on earth visual

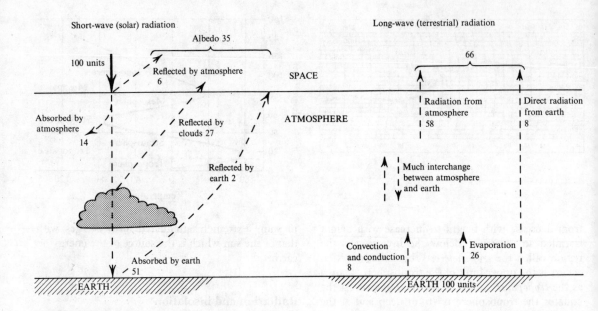

confirmation of the earth's albedo is possible on clear nights with a new moon, when it is sometimes possible to see 'the new moon in the old moon's arms'. The crescent moon is seen by the moonshine which is sunshine reflected from the moon's surface. The remainder of the moon's disc is seen faintly by the reflected earthshine, which is itself the earth's albedo or reflected sunshine. The best conditions for observing the old moon in the new moon's arms are usually found in ridges of high pressure between the depressions in a family (see Fig. 61). Rain in depressions washes impurities from the atmosphere; the ridge provides the cloudless night skies. This is the reason why the phenomenon has long been looked upon as a harbinger of unsettled weather.

47% of the sun's energy or *insolation* (*in*coming *sol*ar radi*ation*) reaches the earth. Some is used in evaporating moisture; some heats the surface of the earth which has an average temperature of about 15°C – higher in some places and lower in others. The earth radiates on an appropriate wavelength which, from the rule already stated, is much longer than the short-wave radiation from the sun. It is therefore spoken of as *long-wave radiation*. Of this

some passes through the atmosphere to space; some is reflected back to the earth by clouds or impurities; some is absorbed by the atmosphere, which also radiates on its wavelength. In this way the clouds and atmosphere act as a kind of screen and long-wave radiation is shuttled backwards and forwards keeping the earth warmer because of the reduced escape. Clouds are very effective screens, as can be demonstrated by comparing temperatures on cloudy nights against clear nights when radiation is very effective.

The total re-radiation to earth from the atmosphere is not quite twice the short-wave radiation received by the earth from the sun. So the atmospheric effect is an important one. Sometimes it is called the greenhouse effect of the atmosphere but, although it achieves a similar result to a greenhouse, the analogy between glass and atmosphere is a very imperfect one.

As over a long period the temperatures of the earth and atmosphere remain constant the heat loss must balance the heat gained. Fig. 7 represents the balance sheet of the heat budget.

3 The general circulation

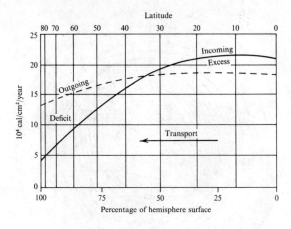

The amount of energy received from the sun varies with the latitude (see Fig. 8). About three times as much arrives at the equator as at the poles. Measurements and calculations show that outgoing radiation is more evenly distributed across the latitudes, so there is a surplus of heat at the equator and a deficit in higher latitudes. As the polar regions are not getting colder and the tropics warmer, equilibrium must be maintained by the transport of heat from low to high latitudes. The largest amount of transfer takes place in middle latitudes which accounts for the unsettled weather which occurs there. The general transport of energy which maintains the balance is called the *general circulation of the atmosphere*. It operates by absorbing heat in the equatorial belt and losing it in the cooler regions towards the poles. To bring about this heat transport the atmosphere employs two different mechanisms.

Subtropical anticyclones
Over the half of the globe between latitudes 30° S. and 30° N. an essential feature is the up-rising air at the thermal equator. This air spreads out northwards and southwards carrying its excess energy with it. Just poleward of the tropics is a region of subsiding air and the return branch of the circulation is the trade winds near the ground. The whole circulation, which is called the trade wind cell or *Hadley cell*, after its discoverer, transfers energy northwards. Closely linked with the subsiding air in the poleward limb of the cell is the belt of subtropical anticyclones which are one of the most permanent features of the circulation. Dry littorals occur where eastern continental coasts intersect the anticyclonic belt. Fed by the dry subsiding air from aloft, these areas are sometimes referred to as the roots of the trade winds. Because of friction between the earth and its atmosphere it is necessary, if the earth's rotation is not to be retarded, for easterly winds to balance westerly winds over the earth as a whole. The easterly trades thus balance the mid-latitude westerlies.

Air in the eastern side of a subtropical anticyclone is not only dry but stable. In its passage westwards over the ocean it is gradually warmed and moistened by evaporation from the warm ocean and in the west side of the high pressure cell it is more unstable so the western sides of the oceanic highs have much moister air which gives generally ample rain. There is in fact a marked contrast between eastern and western coastal areas in this latitude.

Western littorals	20–30°	*Eastern littorals*
e.g. West Indies,		e.g. Namib, Sahara,
China Seas		N. Chile
Onshore winds		Offshore winds
Sea temperatures high		Cold ocean currents
Air generally unstable		Upper air subsidence
Ample rainfall		Persistent drought
Cumulus cloud dominant		Stratus cloud at night
Tropical storms in season		

The upper westerlies
A second mechanism is at work over the other two quarters of the globe north and south of the 30° latitude lines. The maps of the circulation at the 500 mbar level are a useful starting point here. Since half the atmosphere is below 500 mbar maps of the circulation at mid-point in the atmosphere should be a useful guide. Figs. 9(a) and (b) show roughly circular contours with lowest values near the poles – 4920 metres over Antarctica and 5240 metres over the Arctic – rising to 5840 metres near the tropics where the slope then flattens. Winds blow along the contours, so this depicts a circumpolar whorl of upper westerlies in each hemisphere. Because the southern hemisphere has about 80% of its surface as ocean it has the more uniform conditions. Intense cooling over the Antarctic continent ensures a large difference in temperature from frigid to torrid zone. These two facts of geography account for the more

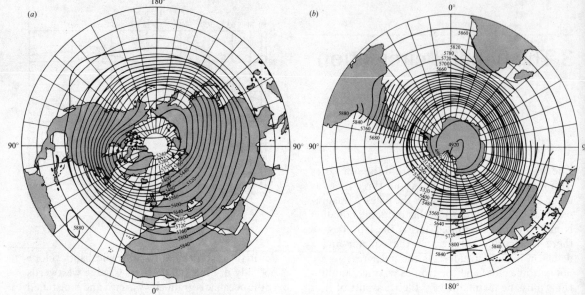

circular, tighter whorl of the southern hemisphere. North of the equator the westerly whorl is less intense. There some energy is carried north in the oceans by the broad warm currents of the Gulf Stream Drift and the Kuro Siwo. (About 25–30% of the heat exchange is affected by the oceans.) In summer the heated continents extend the heat source much further north. In winter when the continents are covered in snow (see Fig. 80) contrasts between the areas north and south of 45° N. are more pronounced and so the winter circulation is much stronger. Calculations have shown that over the year as a whole the momentum of the westerlies in the southern hemisphere is 1.5 times that of the northern hemisphere westerlies. In the northern summer this ratio is increased to 4 to 1, but in the northern winter it is reduced to 0.7 to 1. The greater momentum of the southern hemisphere westerlies has important effects on world climate including the asymmetrical position of the inter-tropical convergence and a preponderance of tropical storms north of the equator.

Figure 9, however, does show average conditions and it is necessary to heed the warning already given concerning climate and average weather. These average maps mask two important facts which need detailed explanation.

Waves in the westerlies

An upper chart for a specific time shows, not a broad westerly stream, but a series of sinuous waves with ridges and troughs (see Fig. 68(b)). Fig. 10 showing the path of a constant altitude balloon circulating at 30° S. to 40° S. indicates the sinu-

osities of the air current at the 12 km level. In 33 days the balloon made three round-the-world trips. The distance along latitude 40° S. is about 30,000 km and so, only allowing a little for the many departures from the latitude, this gives 3000 km per day or an average speed of 120 km/h.

If a smooth flow of air is disturbed, the small departures become accentuated and eventually give rise to long waves of varying amplitude. Days 5 to 7, 20 to 23, and 23 to 25 on Fig. 10 are good examples of waves. An American meteorologist, Rossby, studied the *long waves* (now often called Rossby waves) in the westerlies; showed their connections with mid-latitude weather and how, to some extent, their movements could be predicted. Two, three, four, five or six wave troughs can often be distinguished in the northern hemisphere: three is a more common pattern in the stronger southern hemisphere current. The waves travel more slowly than the winds blowing through them; sometimes they remain stationary and even retreat westwards. The north–south movement of air through the waves helps to carry warm air polewards and cold air equatorwards, thus aiding the global heat transfer. Travelling through a wave trough, air has excess momentum on the inside of the bend and deepens depressions. Conversely a ridge strengthens anticyclonic circulations. These very important systems are explained in more detail later.

Even on the average chart (Fig. 9) deformations from a circular pattern are evident, especially in the northern hemisphere, where two troughs stand out clearly; one over Labrador and the eastern part of North America, and the other over the east of Asia. Since they show up on the average chart they must

10 Trajectory of balloon launched from New Zealand on 30 March 1966 and drifting at the altitude of 12 km. Position plotted at daily intervals.

12 Diagram of the polar-front jet stream

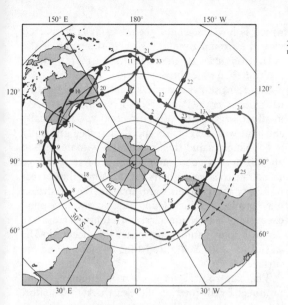

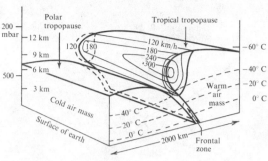

13 Approximate location of the subtropical jet stream (solid line) and area of activity of the polar-front jet stream (shaded) in the northern-hemisphere winter (after Riehl)

11 Disturbance of west-east current of air in the northern hemisphere by a mountain barrier

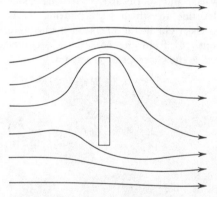

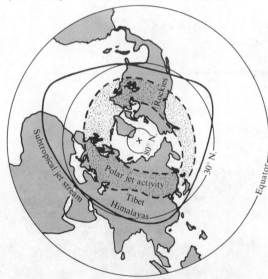

be preferred positions. They are caused by the disturbance of the west wind flow by the Rocky Mountain system in North America and by the mountains of Asia. Fig. 11 illustrates deformation by an orographic barrier. In winter the troughs are emphasised by the continental cold while the warmer North Atlantic and Pacific oceans help to accentuate the intervening ridges which can be distinguished over Spitzbergen and Alaska. Waves are thus initiated by mountain barriers and by broad thermal patterns. The dynamics of air flow preserves, amplifies, and sometimes replicates the waves. The average flow in the southern hemisphere is less deformed, but Fig. 9 shows a shallow trough east of the Andes, a slight ridge south of the Indian Ocean, a trough where the Antarctic coast reaches the polar circle and a broader ridge south of the Pacific.

Jet streams

The average chart shows contours evenly spaced, indicating a smooth gradient for westerly winds. Maps for individual occasions show the gradient concentrated into narrow bands where wind speeds are very high. Upper air soundings have recorded winds of over 400 km/h and winds of over 200 km/h are not uncommon. These bands of very strong winds are known as *jet streams*. A jet stream is a ribbon of air normally thousands of kilometres long; hundreds of kilometres wide, and a few kilometres deep with a lower limit of speed of about 120 km/h. Often the core speed is very much greater than this (see Fig. 12). In the troposphere there are two main jet streams (see Fig. 14). Both are situated just below the tropopause. They are: (1) the subtropical jet stream at about 200 mbar (at the poleward limit of the Hadley cells); (2) the mid-latitude jet-stream,

14 Vertical wind profile through jet stream at Seattle
(48°N. 123°W.), 0300 GMT, 21 April 1958

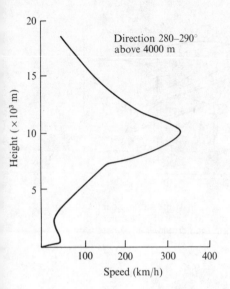

Direction 280–290°
above 4000 m

which is lower at about 300 mbar. It is connected
with the frontal zone and so gets the name of the
polar-front jet. Fig. 13 shows it has a more variable
position than the subtropical jet.

Figure 14 shows a vertical wind profile through a
jet stream over the Pacific coast of North America. As
in all jet streams the wind direction was practically
constant with height. Its speed increased gradually
to 8000 metres and then more rapidly to its peak of
260 km/h at 11,000 metres. The decrease above
12,000 metres was equally rapid. When the correct
humidity requirements are present, the jet stream
may produce cirrus cloud. Fig. 15 shows a good
example of jet-stream cirrus stretching across the
sky in a long band. Most often this occurs in the
warm air aloft well ahead (800 to 900 km) of the
surface front.

The subtropical jet stream is thought to be
produced by the rotation of the earth. At the equator,
the atmosphere which is rotating with the earth has
its greatest angular velocity. Rising air spreading
out northwards and southwards is moving faster
than the latitude to which it is blowing. It is
deflected to the right in the northern hemisphere
and at about 30° from the equator it becomes
concentrated as the subtropical jet stream.

15 Jet stream cirrus (note condensation trail not
extending into the drier air)

16 Temperature distribution at 500 mbar for the
northern hemisphere at 0300 GMT, 6 February 1952
(after Bradbury and Palmen)

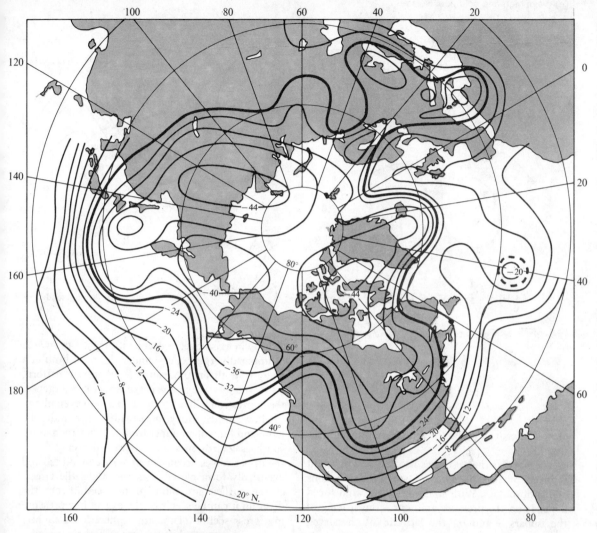

▬▬▬▬▬ Approximate southern limit of polar air (including frontal zone)

The polar front jet is produced by a temperature difference and is closely related to the polar front.

Figure 16 shows the temperature distribution at 500 mbar on 6 February 1952. Temperature varies from $-44°$ C over the Canadian archipelago to $-4°$ C at about $20°$ N.: a difference of about $40°$ C between polar and equatorial regions. The main concentration of isotherms makes a sinuous band from Hudson Bay swinging south of Iceland and north over the Barents Sea before dipping south over Scandinavia and central Europe into the Mediterranean. North of the Caspian Sea it is less definite but shows up again over southern Siberia across northern Japan and the Aleutians. It crosses the Canadian coast in British Columbia and shows a typical ridging in the lee of the orographic barrier of the western Cordillera (cf. Fig. 11). This band separates polar and tropical air: it is the polar front. At height winds blow along the isotherms and closely packed isotherms indicate strong winds. At higher levels the speed will reach jet-stream strength. The cold air over the Mediterranean is only joined by a narrow neck to the main cold air. It is easy to visualise a separate cold pool being formed. This has happened over the Atlantic where a cold 'pool' is centred at about $35°$ N. $35°$ W. Gradually this cold pool will be warmed. This is one way in which polar cold is transferred southwards. Conversely, over the British Isles and Scandinavia warm air is being pushed northwards. The jet stream can be compared with a meandering river. Its meanders become more convoluted and eventually cut off cold lows or warm highs. Fig. 17 shows this diagramatically. It is a mechanism for the transfer

17 Mid-latitude jet stream initiating cut-off cold lows and warm highs in the upper troposphere: (a) long waves; (b) long waves becoming more convoluted; (c) warm high and cold pool isolated

18 Relationship between upper air flow and surface pressure systems in mid-latitudes

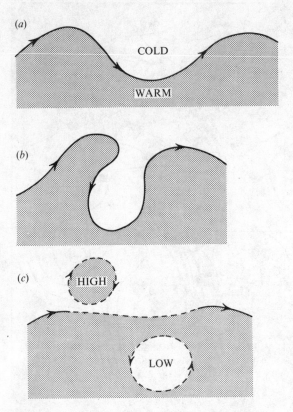

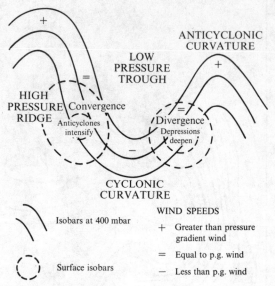

of heat in order to keep the global equilibrium (see Fig. 8).

The waves in the westerlies also have an important influence on the positions of surface depressions and anticyclones. With straight isobars wind speed depends on the pressure gradient – the spacing of the isobars – and on the latitude. With curved isobars, as in a depression, anticyclone, trough or ridge, the centrifugal force has also to be taken into account. For anticyclonic curvature it acts with the pressure force and winds are stronger than the isobar spacing would imply. For cyclonic curvature the opposite is true and winds are lighter than the isobar spacing suggests. Fig. 18 shows equally spaced isobars in the upper atmosphere in a wave situation with a well marked trough and ridge. It indicates positions in which the wind speed is equal to, less than or greater than the pressure gradient wind. In a steady state, air will tend to become concentrated where the speed slackens in the left-hand limb of the trough. This is called convergence of air. Conversely divergence occurs in the right-hand limb where the speed is accelerating. If there is divergence aloft there is usually convergence below. Convergence implies inflow and rising air. Hence depressions tend to form, or if they are

already in existence, to deepen if they are in an area of upper divergence, e.g. under the right limb of a trough. With an unfavourable upper air pattern, e.g. a young wave in an area of upper convergence, the depression will fail to develop or will even fill up. Shallow surface anticyclones readily form under the left-hand limb of an upper trough. They then move south-eastwards in the north-west current. When the upper waves become even more convoluted and eventually form closed lows and highs the effects at the surface are more pronounced. From the diagram it can be seen that the axis of low pressure in a cross section of the atmosphere leans slightly to the north-west while the axis of highest pressure leans to the south-west.

To facilitate this treatment of convergence and divergence a very simple case has been described. Usually, of course, the atmosphere is not in a steady state, nor are isobars parallel for long. But even in the actual complicated atmosphere these processes explain part of the development.

Zonal index

The strength of the westerlies between 35° N. and 55° N. is called the zonal index. Along any meridian a large positive difference of pressure between 35° N. and 55° N. will give a strong gradient for westerly winds. This is a *latitudinal flow*. When this occurs the zonal index is said to be high. Over the British Isles high zonal conditions bring unsettled weather with troughs of low pressure and ridges of high

19 Cross-section showing planetary circulation

20 Average mean sea level pressure in different latitudes (repeated over the poles to show the small polar anticyclones)

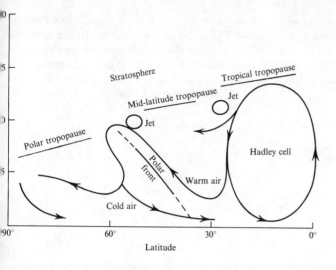

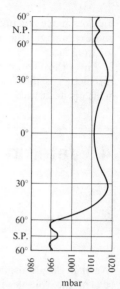

pressure moving in quick succession across the country. As already mentioned Rossby waves develop in the dominantly westerly pattern. The waves continue to develop and as the oscillations become larger air streams tend to blow along the meridians rather than latitudinally. This is called *a meridional flow*. With such meridional flow the pressure difference along a meridian will tend to be small so the zonal index is said to be low. Eventually cold lows to the south and warm highs to the north give a cellular pattern. The warm anticyclones are especially stable and remain stationary while other weather systems move round them. Because of this they are termed blocking anticyclones.

The zonal index varies in an indefinite cycle which is by no means completely understood. Variations in the zonal cycle give different characters to seasons and spells of favourable or unfavourable weather develop under low zonal conditions when the latitudinal flow is least evident.

Low zonal conditions are much less common in the middle latitudes in the southern hemisphere where the westerlies are stronger (see Fig. 9). The northern hemisphere, more complicated by the geographical contrasts of land and sea, develops low zonal conditions more readily. During low zonal conditions there is greater interchange of heat and cold between polar and tropical regions. In chapter 9 some examples of high zonal and low zonal weather situations are discussed.

The general circulation – more detail

Figure 19 shows the meridional circulation in the troposphere. Notice the broken tropopause with the stratosphere above. No longer is the tropopause regarded as a complete lid. Breaks over the two jet streams allow some mixing of tropospherical and stratospherical air. In the winter season the polar stratosphere is hard to identify. At this season of continuous polar darkness the polar stratosphere is cut off from the sun's radiation and becomes very cold. The difference in temperature as compared with the stratosphere further south sets up a strong wind – the polar-night jet stream – which has some influence on the polar troposphere. The polar-night jet is a winter phenomenon found in alternate seasons at both poles.

Figure 19 shows three circulation cells:

(1) The tropical Hadley cell already discussed.

(2) The polar front cell which is sometimes called the *Ferrel cell*. Warm air is shown climbing the polar front and breaking through near the tropopause. The upper part of the polar front is pecked to show this is happening just as it is pecked below to indicate polar outbreaks at low latitudes. The polar front is more continuous and striking in the middle troposphere (see Fig. 16). The main heat exchange occurs aloft and below.

(3) The polar regions have a less well-defined circulation cell. The curve of mean sea level pressure round the earth through the poles (Fig. 20) shows the broad equatorial trough; the subtropical highs in both hemispheres; the mid-latitude trough which is very much more marked in the southern hemisphere, and the high pressure at both poles. The polar anticyclones are far from being permanent features on surface pressure charts but cold

air does move equatorwards off the icy surfaces of these regions and these outflows are fed by subsiding air from aloft.

The other noteworthy feature in the general circulation is the monsoon of the Asian continent which is dealt with in chapter 7.

The general circulation is a remarkable mechanism. Throughout the book more detailed reference will be made to the general concepts introduced in this chapter. It is suggested that this chapter should be read first and then worked over again last in the light of fuller knowledge from later chapters.

4 Water in the atmosphere

Water in the atmosphere, though forming less than 5% of the volume, is mainly responsible for the varieties of weather experienced in the different climatic regions. The bulk of atmospheric water is in the form of vapour which enters the atmosphere by evaporation from sea and land surfaces and by transpiration from plants. The combined total of evaporation and transpiration from a part of the earth's surface is termed *evapotranspiration*.

The main sources of water vapour are the oceanic areas between about 40° N. and 40° S., where the high temperatures experienced during the whole or greater part of the year facilitate evaporation. However, evaporation totals are relatively low where cold ocean currents, such as the Peruvian, have a marked lowering effect on air temperature. For the oceans as a whole the maximum average evaporation amounts to 1150 mm at latitudes 20° N. and 20° S., the greatest totals occurring where subtropical anticyclones, with their prevalence of clear skies, are dominant.

In higher latitudes evaporation from the oceans is greatest where ocean currents such as the North Atlantic Drift have transported warm waters polewards and so raised the temperature, and the moisture-holding capacity, of the air over them. Conversely the presence of cold currents, such as the Labrador current, lowers the amount of evaporation. Over land evaporation is least in those areas with lowest annual rainfall or temperatures.

Humidity

Each of the gases in the atmosphere contributes to the total pressure, i.e. each has a *partial pressure*. Thus, at sea level, nitrogen has a partial pressure of 750 mbars, oxygen 230 mbars and water vapour between 5 and 30 mbars.

It is this partial pressure of water vapour which is used to measure the humidity of the air. When the air is unsaturated the vapour pressure is designated by the symbol e, when saturated by e_s. Values of e_s at selected temperatures are: 6.1 mbar at 0° C; 12.3 mbar at 10° C; 23.3 mbar at 20° C.

The *relative humidity* of the air at a given temperature, i.e. the amount of water vapour in the air expressed as a percentage of the amount present when the air is saturated, is given by the formula:

$$\text{Relative humidity} = \frac{e \text{ (measured)}}{e_s \text{ (known)}} \times 100$$

The vapour pressure is usually measured indirectly using two thermometers mounted in a Stevenson screen. One records the air temperature; the other, with its bulb enclosed in muslin kept moist by distilled water, records the *wet bulb temperature*, i.e. the lowest temperature to which a sample of air can be cooled by evaporating water into it at constant pressure. Unless the air is saturated the wet bulb temperature is lower than the air temperature, since latent heat is absorbed in the process of evaporation. From the dry and wet bulb temperatures the vapour pressure can be calculated. In practice, tables are used to find the relative humidity.

The temperature to which unsaturated air must be cooled at constant pressure for it to become saturated is termed the *dewpoint*. This is lower than the wet bulb temperature and in a sample of unsaturated air remains constant at a given pressure whatever the temperature. The dewpoint is used to estimate the possibility of condensation by cooling.

The *absolute humidity* of the air, i.e. the mass of water vapour contained in a unit volume of saturated air, varies with temperature (Fig. 21). From this graph the amount of water vapour which will condense when the air is cooled can be calculated. Thus 5.5 gm/m^3 will condense if air is cooled from $25°$ C to $20°$ C, 3 gm/m^3 with a decrease from $15°$ C to $10°$ C and 1 gm/m^3 with a fall from $-5°$ C to $-10°$ C. These figures are representative of equatorial, temperate and polar environments and explain some of the rainfall differences between these regions. They also explain why convectional summer storms may be heavy in middle latitudes.

Heat exchanges

Water vapour, liquid droplets and ice crystals co-exist in the atmosphere and changes from one state to another are accompanied by exchanges of heat with the surrounding air as shown below:

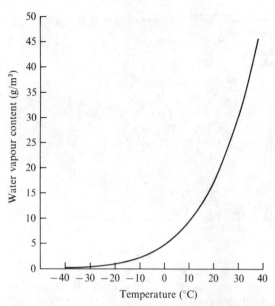

Absorption of heat from the air

Solid

Melting ⟋ ⟍ Sublimation

Liquid ⟶ Vapour

Evaporation

Liberation of heat to the air

Solid

Freezing ⟋ ⟍ Deposition

Liquid ⟵ Vapour

Condensation

Evaporation and condensation are especially important processes.

Condensation nuclei

Laboratory experiments have shown that the water vapour in clean air only condenses when the relative humidity is *over* 100%, i.e. the air is supersaturated. This is because the fastest moving molecules quickly escape from the extremely minute droplets formed by condensation at relative humidities up to 100%, so causing immediate evaporation. Only relatively large droplets with a radius of over 0.0001 mm (0.1 μm) are stable when the air is saturated (R.H. = 100%). The atmosphere, however, contains microscopic particles of impurities, such as common salt, sulphur dioxide and dust on which condensation takes place when the air becomes saturated, and in many instances before saturation is reached. Such particles are called *condensation nuclei*. Salt particles originate from the evaporation of sea-spray, dust from volcanic explosions or wind-blown soil (see dust from the Sahara p. 84) and sulphur from the combustion of coal and oil. Once condensation has begun on a nucleus the droplets grow, rapidly attaining sizes up to 0.05 mm (50 μm) radius. Even the larger droplets formed by condensation are little affected by the earth's gravity and remain suspended in the atmosphere collectively making up clouds or fog.

Freezing nuclei

It has been found that cloud droplets do not automatically freeze until the temperature falls below $-36°$ c. The analysis of snow crystals shows that they include minute particles of insoluble substances, especially certain clay minerals and it has been deduced that these must be present in the atmosphere for freezing to occur. Such substances are called *freezing nuclei*. When they come in contact with a droplet the water from this spreads over them and, if the temperature is below $-10°$ c, immediately freezes. Air which is saturated with respect to water is supersaturated with respect to ice. Once ice crystals have formed further condensation takes place on them. This causes the air in their vicinity to become unsaturated and nearby water droplets to evaporate until the air is again saturated, when further condensation takes place on the ice crystals. Snowflakes grow by this process and eventually become large enough to fall out of the clouds to give precipitation.

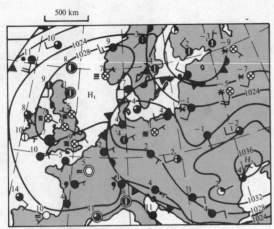

500 km

H₁ Warm anticyclone   1200 GMT   28. 11. 58
H₂ Cold anticyclone

Precipitation also results from the growth of cloud droplets by collision. In middle and high latitudes the freezing process is probably the dominant one; in low latitudes the collision process is important.

Condensation processes

There are three ways in which the air can be cooled and condensation result.

(a) *Advection cooling.* This takes place when air in contact with the earth's surface flows from a warmer to a colder area. If it is cooled to its dew point *advection fog* or *low stratus* cloud results. This happens on the western coastal margins of Europe in winter when a mild, moist, south-westerly flow of air from the Atlantic passes over the colder land surface. The incidence is most pronounced when the influx follows a spell of cold weather during which the ground temperatures have fallen below normal. Fog and low stratus form over the cool North Sea when there is an easterly or south-easterly flow of warm air from the continent of Europe. This gives the North Sea haars along the eastern coast of Britain. Fig. 26 shows an example of this in east Scotland. Two other areas with a high incidence of advection fog are the Grand Banks of Newfoundland and the Californian coast. In both cases it is caused by warm moist air flowing over the cold waters of the Labrador and Californian currents (Fig. 77).

(b) *Cooling by radiation.* The normal daily rise in temperature between dawn and c. 1400 hours is due to the excess of incoming solar radiation over outgoing terrestrial radiation. After c. 1400 hours temperatures start to fall since outgoing radiation from the ground is greater than insolation. The rise in temperature in the early part of the day is accompanied by a decrease in relative humidity and the falling temperatures at night by an increase. The most pronounced night-time cooling of the air near the ground takes place under clear sky conditions when the wind is so light that turbulent mixing with the layers above does not take place. If the stagnating air is cooled to its dewpoint either *dew* or *radiation fog* forms. Warm anticyclones in which the surface layers of air have a high humidity are particularly conducive to fog formation in middle latitudes during the long nights of the winter half of the year (Fig. 22).

Dew forms when the wind speed at 2 m is less than 10 km/h and the air close to the ground is cooled to its dewpoint. It forms on the leaves of plants, especially grass; the condensation taking place on condensation nuclei on the leaves. The water vapour which condenses comes partly from the air and partly by evaporation from the soil. The deposition of dew results in a decrease in the absolute humidity of the air and, consequently, a lowering of the dewpoint.

Radiation fog forms when a much greater depth of moist air is cooled to its dewpoint by outgoing radiation. For fog to form a light wind of 3–10 km/h is required to give sufficient turbulence to mix the air chilled by contact with the ground with that above. Since fog formation is normally associated with some deposition of dew it will only occur if the air temperature falls more rapidly than the dewpoint and the *fog point*, several degrees below the original dewpoint of the air, is reached. Radiation cooling from condensed droplets as well as from the ground is a factor leading to the final stage of cooling of the air and the attainment of the R.H. of 100% necessary for fog formation. Once a layer of fog has formed continued radiation from the top of the fog results in the development of a temperature inversion. Radiation fogs are particularly found in valley bottoms since chilled air increases in density and sinks down valley sides. Thick fogs may persist for several days. Fog dispersal results either from a sufficient intensity of day-time insolation to cause evaporation or an increase in wind speed to mix the fog layer with the drier air above.

(c) *Cooling by ascent.* When air ascends it undergoes expansion since atmospheric pressure decreases

with altitude. The expenditure of energy which occurs in the expansion process results in a decrease of temperature. Such a temperature change involving no addition or subtraction of heat is termed *adiabatic*. The rate of decrease which occurs when air rises under these conditions is called the *adiabatic lapse rate*. In unsaturated air the rate of cooling or the *Dry Adiabatic Lapse Rate* (D.A.L.R.) is $10°$ c/km. When the dewpoint is reached condensation results in cloud formation. Because of the liberation of latent heat of condensation the rate of cooling of saturated air or the *Saturated Adiabatic Lapse Rate* (S.A.L.R.) is less than that of dry air. Near sea level it is $6°$c/km but gradually approaches the D.A.L.R. as altitude increases and the moisture-holding capacity of the air falls with the decrease in temperature.

The ascent of air leading to cloud formation arises in a number of ways but gives rise to two major groups of cloud types.

(*a*) Heap or cumuliform clouds (e.g. Fig. 23) which form when bubbles of air can ascend freely. The atmosphere under these conditions is *unstable*.

(*b*) Layers of cloud which form when the air is stable but is forced to ascend, e.g. on meeting rising

ground or a mass of denser air.

Cloud forms in unstable air

The stability or instability of a particular mass of air depends on its temperature and humidity at different levels. Information about this, which allows the *Environmental Lapse Rate* to be drawn (Fig. 24), is of vital importance to weather forecasters. The information is now obtained from radio-sondes. These are balloons with electrical instruments, which record the pressure, temperature and humidity of the atmosphere, attached to them. Radio-sondes ascend to great heights. A ground station receives signals from the radio transmitter carried by the sonde and decodes them. The dry bulb and dewpoint temperatures at the different pressure levels are plotted and the graphs interpreted.

Figure 24 shows one form of *adiabatic chart* on which upper air observations are plotted. The horizontal lines are pressure levels which are directly related to height. The vertical lines are isotherms with the highest temperatures on the right. Sloping upwards from right to left as pecked lines are the dry adiabats which show the rate at which rising

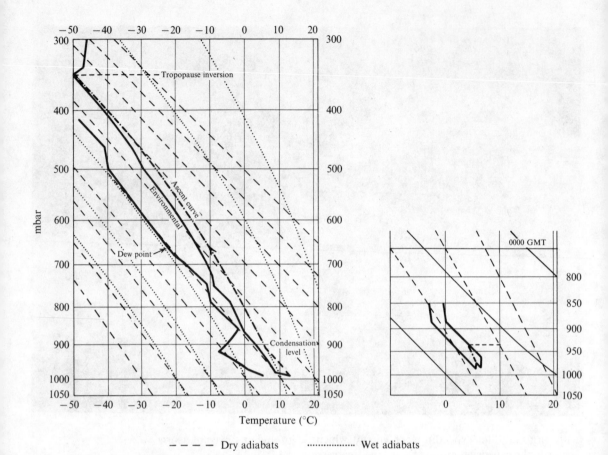

– – – – Dry adiabats ················· Wet adiabats

unsaturated air cools. The dotted lines which rise more steeply than the dry adiabats, indicating a slower rate of cooling, are the saturated adiabats.

Plotted on Fig. 24 is the ascent for 1200 GMT, 30 April, at Aughton, Lancashire. The air temperature at ground level was 13° C and the dewpoint 5° C. If a parcel of this air had risen it would cool at the D.A.L.R. until it reached saturation at the condensation level, which was about 900 mbar (800 m). At this level the dewpoint was still 5° C, the air saturated with moisture and cloud formed. Any further uplift caused the air to cool at the S.A.L.R. This is shown as the ascent curve on Fig. 24. It will be seen that air following the ascent curve was warmer and therefore lighter than that of the environment and would continue to rise until the ascent curve met the environment curve. This occurred at 345 mbar (8000 m) which, in this case, was the tropopause (cf. Fig. 42, where the tropopause near the equator is at 16,000 m). The large *cumulonimbus* clouds (Fig. 25) which formed as a result were up to 7000 m thick and gave the heavy showers and thunderstorms shown on the 1800

GMT chart (Fig. 26). Instability was further favoured by the fact that the British Isles were covered by a shallow depression centred over East Anglia and the convergence of low level unstable air into the centre of this depression favoured the persistence of the showery conditions into the evening.

The inset to Fig. 24 shows the environment and dewpoint curves up to 850 mbar at 0000 GMT on the same day. At this time the surface pressure was 987 mbar, the air temperature 6° C and the dewpoint 5° C. If a parcel of this air had risen, e.g. up a hillside, it would have cooled at the D.A.L.R. until the dewpoint was reached at about 1000 mbar (100 m) above the surface. Further ascent would have resulted in cooling at the S.A.L.R. The rising air would have remained colder and denser than the environment up to 930 mbar (570 m) and would therefore sink back to its original level if no longer forced to rise. Air with such properties is *stable*.

If the parcel of air had been lifted above 570 m it would have become warmer than the environment and would have risen freely to the 1370 m (850 mbar)

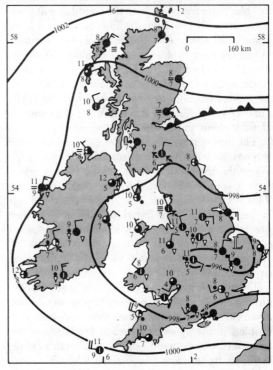

level, by which time it would have cooled to the temperature of the environment and no further unforced ascent would have been possible. Between 570 m and 1370 m the air was *unstable* and from ground level to 1370 m *conditionally unstable*, i.e. convection currents would develop once the air was lifted to the level where it became warmer than the environment.

Unstable air masses often become stable during the night due to radiational cooling. After sunrise the sky is often clear and insolation warms the surface layers slowly or quickly, depending on the time of year. With increase of temperature the lower layers lose their stability, air begins to rise and cumulus clouds to form. At first they are small but with increased insolation the instability extends to higher levels and, in favourable conditions, cumulonimbus clouds form.

Figure 27 illustrates the process leading to the formation of 'thermals' and cumulus clouds. Thermals form most easily where temperatures are raised most quickly by insolation, e.g. over rock outcrops or dry soil or large areas of concrete. A thin film of air in contact with the heated patch of ground expands and within a short time forms a bubble of warm air which breaks away and rises

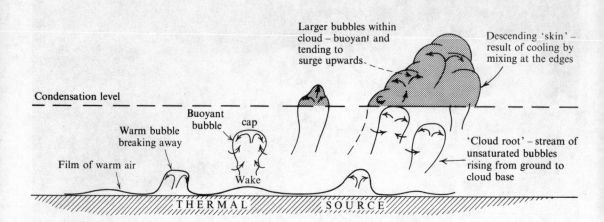

Larger bubbles within cloud – buoyant and tending to surge upwards.

Descending 'skin' – result of cooling by mixing at the edges

Condensation level

Buoyant bubble

cap

Warm bubble breaking away

Film of warm air

Wake

'Cloud root' – stream of unsaturated bubbles rising from ground to cloud base

THERMAL   SOURCE

through the surrounding cooler air. A cumulus cloud begins to form when the bubble top rises above condensation level. If the air is unstable for only a small height above condensation level small, or fair-weather, cumulus result (Fig. 23). When the air is unstable to great heights and a succession of bubbles enter the cloud base the cumulus will grow both vertically and laterally. Fig. 28 shows a large cumulus in which several bubbles can be identified. When the supply of additional bubbles ceases the cloud begins to dissipate. This is very evident when cumulus clouds evaporate over land in the evening as thermal development ceases or when they drift out over a cold sea. The maximum life of a cumulus cloud is about 30 minutes.

The tops of large cumulus clouds often reach the tropopause when the air is unstable to that level. When the temperature in the cloud tops is below $-10°$ C some of the water droplets begin to freeze. This is called *glaciation*. Freezing of the cloud top is one cause of precipitation, (see p. 15) and precipitating cumulus clouds are called *cumulonimbus*. On reaching the tropopause the tops of cumulonimbus spread out to form an anvil of *cirrus* cloud (Fig. 25).

Thunderstorms are associated with large cumulonimbus clouds in which there are very strong up and down currents (Fig. 29). When super-cooled water droplets in the up-currents reach levels where the temperature is between $-10°$ C and $-30°$ C some begin to freeze from the surface inwards. Their colder outer surfaces become positively charged

with H^+ ions and the warmer cores negatively charged with OH^- ions. As freezing progresses inwards the resultant expansion causes the soft hail stones so formed to shatter. The positively charged splinters from the surface shells are carried upwards whilst the negatively charged larger cores fall down. Coincident with the separation of electrical charges within the cumulonimbus clouds the ground beneath them acquires a positively induced charge, instead of the normal negative one. When the electrical charge between the upper and lower parts of the cumulonimbus clouds and between the clouds and the ground is sufficiently strong to overcome the resistance of the air a discharge takes place, causing a lightning flash.

The precipitation associated with thunderstorms is in the form of very large raindrops or hail stones which are kept in suspension, or are carried repeatedly upwards by the strong up-currents within the cloud, until they have grown to a sufficient size to continue on a downward path. When this stage is reached a sudden fall-out occurs and causes a down-draught of cold air which spreads out below the storm (Fig. 29).

The cold down-draughts which develop with the onset of intense precipitation eventually become dominant. As they spread below the cloud they cut off the supply of warm, moist air necessary for continued convection. The cumulonimbus cloud then begins to dissipate, leaving irregular masses of cloud at various levels. The cirrus of the anvil top persists longest.

28 Large cumulus – active on top left, decaying on top right
29 Vertical air currents and distribution of electrical charges in a vigorous cumulonimbus

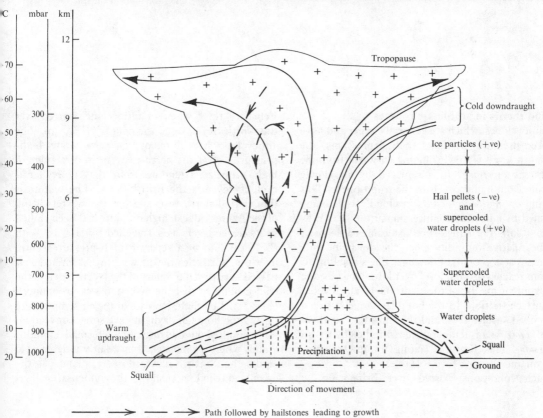

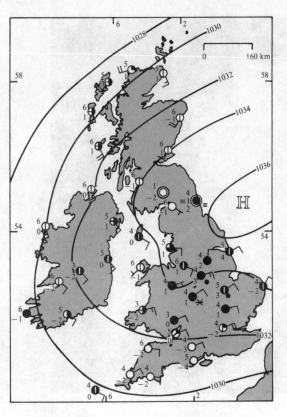

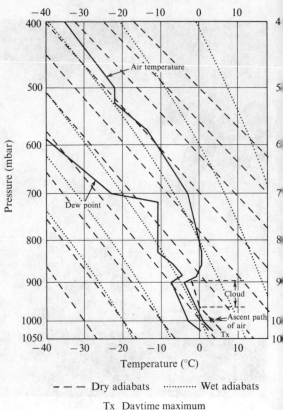

– – – Dry adiabats   ·········· Wet adiabats

Tx Daytime maximum

Cloud forms in stable air masses

In anticyclones, which are generally larger and have weaker pressure gradients than depressions, the air at low levels is slowly moving outwards from the centre or *diverging*. It is replaced by air slowly sinking from higher levels in the troposphere. This sinking or *subsidence* results in cloudless air being warmed by adiabatic compression at the D.A.L.R. of 10° c/km. The subsidence also causes a decrease in the relative humidity since the dewpoint only increases at 1.7° c/km of descent. The subsiding air seldom reaches ground level over lowland areas due to turbulent mixing of the air in the layers affected by surface friction. In the boundary zone between the mixed surface layer and the subsiding air there is an *increase* of temperature with height or an *inversion*. This forms a ceiling to upward convection and the air above is stable.

Such conditions existed over Britain on 27 February 1968, when an anticyclone centred over the North Sea covered the country (Fig. 30). The anticyclone had developed as a separate high-pressure cell at the north-eastern tip of a ridge of high pressure extending from the Azores anticyclone towards the British Isles. The warm air transported north-eastwards on the western flank of the Azores ridge of high pressure had been chilled in its lower layers as it travelled polewards. With the formation of a separate high-pressure centre over the North Sea the air was further chilled as it passed over the cold waters of the North Sea and the cold ground so that the inversion was pronounced (Fig. 31). The slight increase in the air temperature during the hours of daylight gave some convection. This resulted in the formation of small cumulus clouds where air moistened by a fairly long passage across the North Sea was flowing inland. As the ascent curve in Fig. 31 shows, the condensation level

33 (*below*) Development of stratus

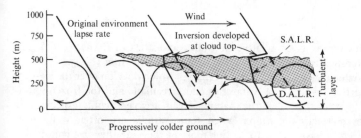

in mid-afternoon was at 960 mbar (700 m) and the upper limit of convection at 895 mbar (1350 m). In such a shallow unstable layer the cumulus clouds spread out at the base of the inversion to form a layer of stratocumulus (Fig. 32). The cloud layer on this occasion was thick enough in places to give occasional light precipitation. Over the southern-most parts of England (Fig. 30) there was a greater difference between the air and dewpoint tempera-tures compared with East Anglia and the Midlands. This and the clear sky conditions reported reflect the drier nature of the air which had followed a short sea passage after leaving the continent.

Stable air is lifted in other ways to cause cloud formation. When moist sea air spreads inland over a cold land surface the resultant increase in turbulence over the land, compared with that over the sea, combined with cooling of the turbulent layer by the ground, results in formation of a layer of *stratus* cloud which extends from condensation level to the base of the inversion at the top of the turbulent layer (Fig. 33).

Other ways in which layer clouds form are dealt with in chapters 5 and 13.

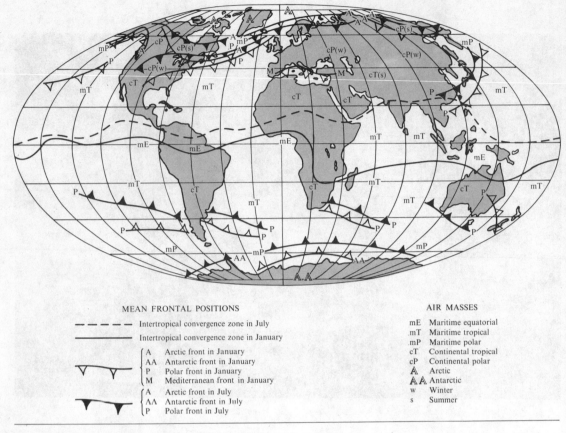

MEAN FRONTAL POSITIONS		AIR MASSES	
– – – – – Intertropical convergence zone in July		mE	Maritime equatorial
───── Intertropical convergence zone in January		mT	Maritime tropical
		mP	Maritime polar
A	Arctic front in January	cT	Continental tropical
AA	Antarctic front in January	cP	Continental polar
▽▽ P	Polar front in January	A	Arctic
M	Mediterranean front in January	AA	Antarctic
A	Arctic front in July	w	Winter
▼▼ ΛΛ	Antarctic front in July	s	Summer
P	Polar front in July		

5 Air masses and frontal systems

An air mass is a large body of air in which there is only a very gradual horizontal change in temperature and humidity at any given height and in which lapse rates are almost uniform. These properties result from the air stagnating over a particular part of the earth's surface for several days or even weeks. This happens particularly in the large, quasi-stationary anticyclones of the subtropical and polar regions throughout the year and in the anticyclones which form over the cold parts of Eurasia and North America in winter. During the period of stagnation the air may be warmed or cooled 10° C or more throughout the whole thickness of the troposphere.

Air masses are designated *tropical* when formed over a warm part of the earth's surface; *polar* or *arctic* when formed over cold parts. Air masses forming over the oceans acquire a high relative humidity, especially in their lower layers, and are called *maritime*; those forming over land have a low relative humidity and are called *continental*. The main source regions are shown in Fig. 34.

Arctic air masses which develop over the ice-caps of Antarctica and Greenland and the frozen waters of the Arctic Ocean are very cold, especially in winter when temperatures are often well below −40° C. For most of the year they are continental (cA) in their humidity characteristics but those developing over the Arctic Ocean in summer, when the sea ice is partially melted, are maritime (mA) in character.

Continental polar (cP) air is similar to cA in the lower layers of the troposphere but somewhat warmer at higher levels. The largest source region is the Eurasian continent east of 20° E. and north of the Himalayan mountain system in winter, and north of 60° N. in summer. In North America the Mackenzie Basin and the Northern Great Plains are the main source regions. There are no sources of cP air in the southern hemisphere.

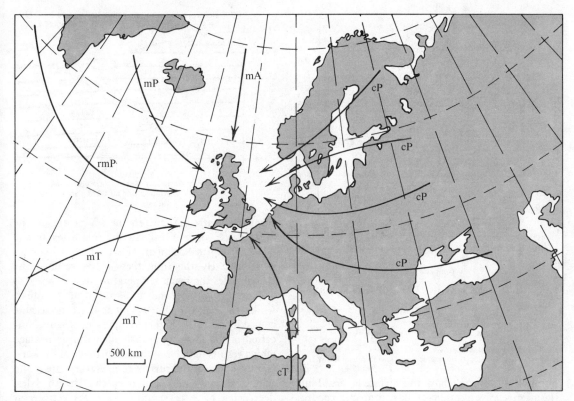

Maritime polar (mP) air originates over the cold waters of the Antarctic Ocean south of *c*. 40° S. and the northern parts of the North Pacific and North Atlantic, especially where cold ocean currents occur. It is also formed in the northern hemisphere from the modification of cP air as it is warmed and moistened after leaving the land and following a long sea track. This is especially so over the oceanic waters to the east of Canada and Siberia.

Continental tropical (cT) air has its source regions over the tropical deserts such as the Sahara throughout the year and over the temperate deserts of the inner parts of Asia in summer.

The main sources of maritime tropical (mT) air are the semi-permanent oceanic subtropical high pressure cells.

Tropical air masses become modified to form equatorial (E) air when they reach and stagnate in the doldrums.

Air masses affecting western Europe
Once a mass of air starts to move away from a source region it undergoes modifications in temperature, humidity and stability. Air which travels towards cooler regions becomes chilled and generally more stable; that moving towards warmer regions becomes warmer and less stable. Continental air undergoes an increase in humidity once it starts to travel over a sea or a large lake; maritime air decreases in humidity when precipitation occurs during a long land track.

The air masses which affect western Europe are shown in Fig. 35, whilst Fig. 36 illustrates the mean temperature changes undergone by mP and mT air masses in winter during their passage towards the British Isles.

Tropical air masses
Maritime tropical air periodically reaches the British Isles and western Europe throughout the year. The graphs of mean temperature conditions in Fig. 36 show that, in mid-winter, the mean surface temperature falls from *c*. 15° C to *c*. 10° C between the Azores source region and the British Isles; that at the 450 mbar level from *c*. −21° C to *c*. −23° C. The stability of the air, which has a lapse rate less than the S.A.L.R., is shown by the convexity of the graphs towards higher temperatures. It is this inherent stability which prevents large-scale convection in mT air and causes the low

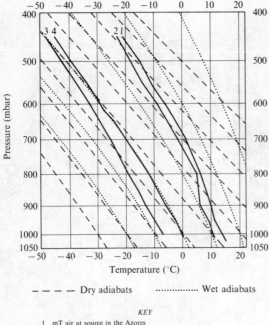

---- Dry adiabats   ·········· Wet adiabats

KEY

1 mT air at source in the Azores
2 mT air over the BIs which has followed a direct track from the Azores
3 mP air at Icelandic source
4 mP air over BIs which has followed a direct track from Iceland

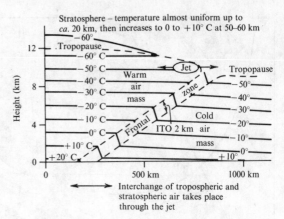

Interchange of tropospheric and stratospheric air takes place through the jet

humidities recorded above the surface layers. Thus the mean relative humidity at the 500 mbar (*c.* 5000 m) level is between 64% and 86% in mid-winter and between 53% and 69% in mid-summer. In the surface layers a R.H. of 85% or over is common in mid-latitudes. Because of the warmth of the mT air however, that reaching the British Isles contains, on average, 2.6 gm/m³ of precipitable water in the 950–450 mbar layer in winter and 1.8 gm/m³ in summer. At mid-summer the mean surface temperatures are 22°C at the source and 19°C over the British Isles, the 450 mbar temperatures being − 12°C and − 16°C respectively.

Continental tropical air from the Sahara source region reaches the British Isles only in summer. At the source region mean temperatures in it are 32°C at the surface and − 12°C at 450 mbar; over the British Isles the corresponding temperatures are 23°C and − 14°C. Although the moisture content of the turbulent surface layers increases in the passage of cT air over the Mediterranean and the English Channel the R.H. is lower than that of mT air. Even so the 450–950 mbar layer contains 1.4 gm/m³ of precipitable water.

Polar air masses

The ascent curves for mP air in Fig. 36 give an indication of the greater instability of this air mass

compared with mT air. In the source region the mean lapse rate of mP air approximates to the S.A.L.R. up to 800 mbar, above which it is somewhat less. By the time the air has reached the British Isles the mean temperature has risen 5°C to 6°C at the surface and 3°C to 4°C at 450 mbar, whilst the curve up to 750 mbar has a noticeable concavity towards higher temperatures – an indication of the increase in instability during passage which results in the frequent occurrence of showers. In mid-summer the surface temperatures are *c.* 8°C at the source and 14°C over the British Isles, decreasing to − 34°C and − 31°C respectively at 450 mbar. The instability of this air causes the upward transportation of heat, through the release of latent heat of condensation, and water vapour. The relative humidities are greater at higher levels than in mT air.

Maritime polar air which has first travelled southwestwards from the Icelandic areas and then followed a long west to east track towards the British Isles is called *returning maritime polar* air (rmP). It is warmer and more stable than direct mP air.

The mean temperatures in cP air reaching the British Isles in winter are − 2°C at the surface and − 43°C at 450 mbar. The humidities are much lower than in mP air and it is consequently more stable. If it has a long passage over the North Sea the moisture content and instability increase to give showers on the eastern side of Britain.

The instability of any air mass increases with cyclonic development which results in ascent and adiabatic cooling. Conversely subsidence and adiabatic warming associated with anticyclonic development lead to an increase in stability.

Fronts and frontal depressions

The average differences in temperature between

(a) Frontal wave stage
(b) Mature stage
(c) Occluding stage
(d) Dying stage

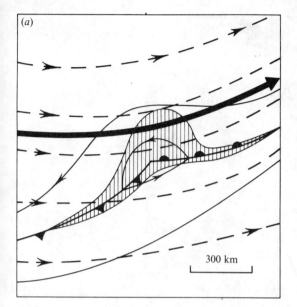

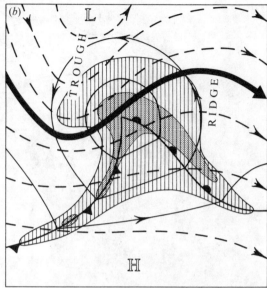

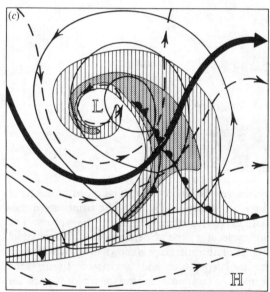

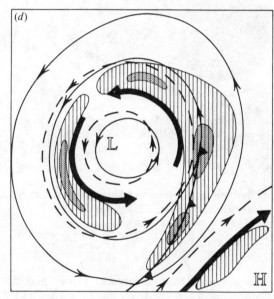

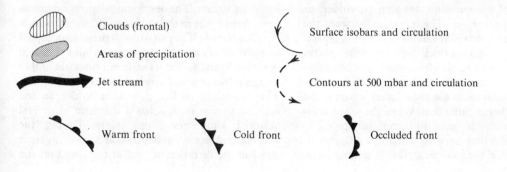

Clouds (frontal)

Areas of precipitation

Jet stream

Surface isobars and circulation

Contours at 500 mbar and circulation

Warm front   Cold front   Occluded front

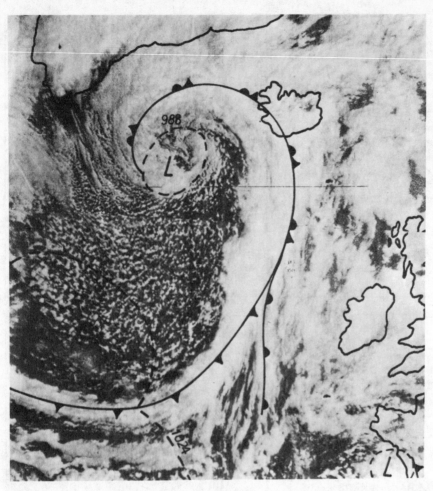

39 Satellite photograph (Nimbus 3), 17 September 1969, showing a decaying depression south-west of Iceland. Note frontal clouds, cumulus behind the cold front and broken layer cloud in the warm air

mT and mP air masses which have followed a direct track towards the British Isles are:

Pressure level	950 mbar	700 mbar	450 mbar
Summer	7°C	12°C	15°C
Winter	12°C	18°C	22°C

These temperature differences cause marked contrasts in density between the air masses and, where low level convergence results following the formation of a depression, the denser, colder air undercuts the warmer. The transition between the two air masses, in which mixing occurs, is a *frontal zone*, usually 1 to 2 km thick (Fig. 37). As the warm air is forced upwards the processes of condensation lead to cloud formation. If the warm air is stable, layers of cloud form in the moist layers within it, the drier levels being cloud free. When the moist air is unstable the clouds are of the cumuliform type.

The frontal zone between polar and tropical air masses is called the *polar front*, that bounding arctic

air masses the *arctic front*. The mean positions of these are shown in Fig. 34. The fronts are quasi-stationary when lying parallel to the isobars. The life cycle of a frontal depression is illustrated in Fig. 38. The initial stage is shown in (*a*) when a small surface depression and associated fronts have formed in response to the onset of upper air divergence leading to convergence in the lower layers (pp. 11–12). The ascending air is removed by the jet stream. The developing depression causes a poleward bulge in the warm air and the formation of a *frontal wave*. The part of the depression occupied by warm air is called the *warm sector*. An increase in the wind speed in the jet stream causes the upper trough to deepen and accelerates the ascent of air. This causes the surface depression to deepen and increase in size until it has a diameter of several hundred kilometres (Fig. 38(*b*)). During the deepening process the depression and warm front travel in the direction of, and at the speed of, the

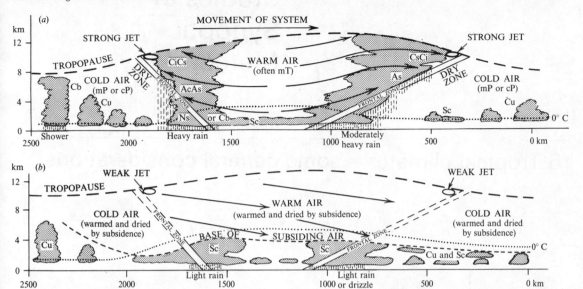

40 Sections through warm sectors of polar-front depressions (after Pedgley)
(a) Ana fronts. Warm air rising relative to frontal surfaces to give thick frontal clouds

(b) Kata fronts. Upper air sinking relative to frontal surfaces and vertical extent of clouds limited by subsidence inversion

wind in the warm sector. The cold front usually travels faster than the warm since the wind speed in the rear of a depression is normally greater than in the warm sector. The cold front consequently starts to catch the warm one up, the process commencing at the tip of the warm sector. When this happens the warm air is lifted above the earth's surface to give an *occluded front* or *occlusion* with cold air ahead of and behind it. If the air behind is the colder the occlusion is *cold*; if it is the warmer the occlusion is *warm*.

The occluding process is associated with the formation of a cut-off low in the upper and middle troposphere in which upper air convergence takes place with the winds spiralling inwards in the form of a vortex. This is picked out by the medium layer pattern (Fig. 38(c)). The satellite photograph (Fig. 39) clearly shows this. Convergence into the upper low is accompanied by a gradual rise in pressure in the surface depression and a decrease in its speed. As the depression becomes quasi-stationary the air circulating round it acquires almost uniform characteristics so that distinctive air masses can no longer be identified and the frontal cloud slowly dissipates (Fig. 38(d)). The decay of fronts is called *frontolysis*.

Figure 40(a) shows a section through the warm sector of a vigorous frontal depression, in which the warm air is rising at all levels relative to the frontal surface. Such frontal systems, called *ana* fronts, give prolonged, heavy precipitation. When the temperature contrasts between the warm and cold air masses are slight, and the upper part of the troposphere is occupied by subsiding, relatively warm, dry air, the frontal clouds do not extend to great heights and any precipitation is light (Fig. 40(b). Fronts of this type, called *kata* fronts, are particularly found when the frontal system is travelling round the periphery of an anticyclone. (For this terminology compare anabatic and katabatic winds, pp. 89–90.)

Inter-tropical convergence zone

The air from the source regions of mT and cT air masses moving towards the equator is usually travelling from colder to warmer regions. This particularly applies to the air which flows equatorwards from the more stable eastern parts of the subtropical high pressure cells of the Atlantic and Pacific. Such air has been cooled in its lower layers in passing over the cold ocean currents (Californian and Peruvian in the Pacific; Canaries and Benguellan in the Atlantic). Moving equatorwards in the tradewind belt it becomes warmer, moister and more unstable. The temperatures, humidity and density of air from the northern and southern hemispheres, even with a wholly oceanic track, are seldom identical. Marked uplift takes place in the convergence zone giving rise to belts of massive Cb clouds and periods of intense precipitation. This is the *inter-tropical front* or *inter-tropical convergence zone*. The convergence is most marked where cT and mT air are involved as over west Africa, particularly in June and July; over southern Asia in July; and over northern Australia in December and January. The positions of the I.T.C.Z. at the solstices are shown in Fig. 34.

Part 2
Studies in
Synoptic
Meteorology

6 Tropical climates – some general considerations

An elementary approach to climate in the tropics, distinguishes three main belts:

(a) *The doldrums*, a narrow belt of light variable winds in the heart of the tropics. This is a region with thick clouds and heavy rain showers. On the pressure chart it shows as a region of low pressure, known as the equatorial trough, or more correctly, as the inter-tropical trough, because it is not always at the equator.

(b) *The horse latitudes*, a region near, or just pole-ward of, the tropics of Cancer and Capricorn, with cloudless or nearly cloudless skies and light winds. Here surface pressure is at a maximum in the subtropical anticyclones.

(c) *The trade wind regions*. These are broad areas characterised by regular, dependable winds blowing from an easterly point. They blow out from the anticyclones towards the equatorial trough.

These three belts in the northern hemisphere can easily be distinguished on Fig. 41, which shows part of the tropical Pacific Ocean. Remember half the area of the surface of the globe lies between 30° N. and 30° S. This chart shows about one-sixteenth of the earth's surface in a region little affected by the differences between land and sea.

At about 30° N. are the cells of the subtropical anticyclone with a central pressure of 1022 to 1025 mbars. Two cells appear in this limited area, though the average map shows only one large anticyclone. Air temperatures are 23° C; clouds are very scattered fair-weather cumulus (see Fig. 23). South of the anticyclonic cells is a broad belt of

easterly winds – the north-east trades. There is some variation in direction but normally there is always an easterly component.

A small trough with its axis from 25° N. 165° E. to 7° N. 153° E. breaks up the regular pattern of the isobars in the trade-wind area. This feature is an easterly wave which will be mentioned again later. In the trades, away from the anticyclone, large cumulus clouds are reported and showers are occurring. South of the trades is the low pressure trough of about 1009 mbars, with its axis near, though slightly north of, the equator. On this chart the inter-tropical trough is only slightly displaced from the equator though in late April the midday sun is overhead at about 13° N. Temperatures average about 28° C. The islands of Eniwetok and Nauru are 700 miles apart but show a pressure difference of only $3\frac{1}{2}$ mbars. In Britain this would be a remarkably low pressure difference but near the equator pressure gradients are never strong because the horizontal component of the deflecting force of the earth's rotation is absent at the equator, and air can flow direct from high to low pressure. Hence pressure differences are hard to sustain.

South of the equatorial trough is the northern edge of the south-east trade-wind belt. Often a line is drawn in the equatorial trough to indicate the separation of air from the north-east trades and air from the south-east trades, and sometimes this line is called the inter-tropical front. On Fig. 41 it would pass between Nauru and Eniwetok. Soundings of the upper air temperatures at these two islands (1100 km apart) were as shown in the table below Fig. 41 on the next page.

41 Synoptic chart, Marshall Islands, Pacific Ocean, 0000 GMT (midday observations), 28 April 1951

42 Upper air diagram, Nauru, (0° 25′N. 166° 0′E.), 28 April 1951

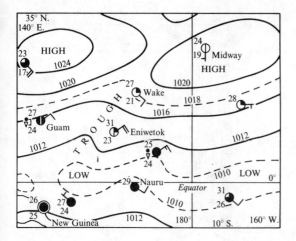

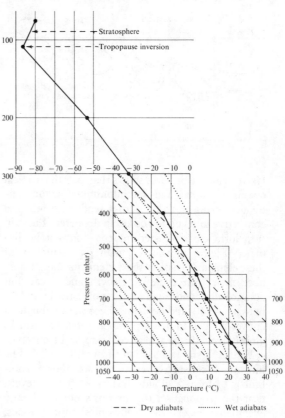

Pressure (mbar)	Nauru (°C)	Eniwetok (°C)
1000	29.0	29.1
900	21.0	21.0
850	18.0	18.1
800	15.4	15.0
700	9.2	9.1
600	3.0	2.8
500	−4.8	−4.8
400	−15.3	−15.1
320	−27.0	−26.0

The most noticeable thing is the similarity of the air temperature at all levels above the two stations. Care must be taken not to think of the inter-tropical front as similar to a polar front. Notice on the sounding (Fig. 42) that the tropopause is reached at a pressure of about 100 mbar which corresponds to about 18,000 m. Above this level temperature increases with height in the stratosphere. In the inter-tropical trough heavy cumulonimbus cloud can tower to this height of about 18 km. It is a zone of convergence where air streams flow together. When this happens air must ascend, and if the air currents are moist, this leads to condensation, clouds and rain. Conversely where there is divergence or out-flowing air, as in an anticyclone there must be subsiding air. The subtropical anticyclones are such regions.

Cloud cover photographs from satellites are now increasing our knowledge of weather over the tropical oceans. Fig. 43 shows the clouds associated with active weather systems over the Indian Ocean in February. Very prominent is the cloud band associated with the inter-tropical convergence zone (I.T.C.Z.) lying from Malagasy to Sumatra. Ana-

43 Clouds (80–100% cover) in the I.T.C.Z., Indian Ocean, 8 February 1967

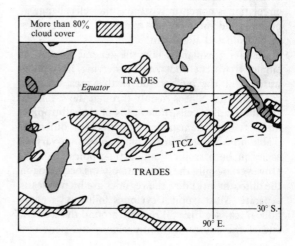

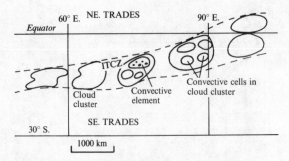

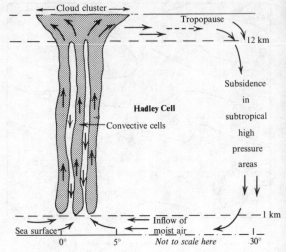

lysis of satellite pictures shows that over the tropical oceans cumulus and cumulonimbus clouds individually up to 10 km across tend to group into convective units up to 100 km in diameter. These in turn form cloud clusters of 100–1000 km in diameter. Figure 44 shows a plan representation of these elements. Usually these elements and clusters have their own cycle of growth and decay but occasionally conditions on the poleward boundary between trades and doldrums become suitable for a convective cell to develop into a tropical storm. In global terms this is a relatively rare occurrence.

Figure 45 shows a section through I.T.C.Z. The individual convective elements are shown combining to form a cloud cluster. At higher levels, near the tropopause, the air flows polewards and eventually feeds the subtropical anticyclones, though some heat is also carried to middle latitudes. In the anticyclones air subsides and is warmed adiabatically. When it reaches the surface it is drier air. This is the horse-latitude belt at about 30° N. On the equatorward side of this belt the trades start as dry winds and pick up moisture over the warm sea becoming moist to a greater and greater depth. Frictional in-flow below cloud level leads to convergence and the cycle continues. This Hadley circulation is not entirely steady: the cells change in position and intensity and the I.T.C.Z. fluctuates its position and alignment.

Chapter 3 explained how the general circulation employs two methods of transferring heat polewards. For the half of the earth's area between 30° N. and 30° S. it is mainly through the I.T.C.Z. and the Hadley cells. Polewards of the subtropical highs the mechanisms of transfer are different. Table 6.1 shows the main mechanisms on various scales in both tropical and extra-tropical regions. However, despite the different systems occurring in the different latitudes, the regimes are by no means separate. Most tropical cyclones following a parabolic track (see Fig. 58) recurve around the western end of the subtropical anticyclones and penetrate

46 Daily positions of cold front penetrating the tropics, 25–30 January 1957

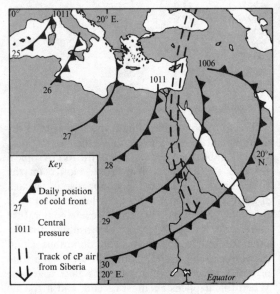

into middle latitudes taking on eventually the characteristics of extra-tropical depressions.

Polar air, modified in its passage through middle latitudes is continually reaching the tropics. One example is shown on Fig. 46, which shows a well marked cold front penetrating into the southern Sudan. As the parent depression moved east from the Gulf of Genoa to Israel in four days it deepened to 1004 mbar. In its easterly position the depression drew in polar continental air from western Siberia and on 5 February Khartoum had a minimum

Table 6.1 *Main dynamical processes in the atmosphere classified by size and time duration*

	SCALE OF DISTANCE			
10,000 km	1000 km	100 km	10 km	0 km
WORLD SCALE	SYNOPTIC SCALE	MESO-SCALE		CONVECTIVE OR SMALL SCALE
Extra-Tropical Long or Rossby waves / Hadley cells or subtropical anticyclones	Depressions / Anticyclones	Fronts Lee waves / Squall lines in troughs		Cumulonimbus showers / Tornadoes
Tropics I.T.C.Z. / Easterly waves	Cloud clusters / Tropical cyclones	Meso-scale convective cells		Convective elements
1000 h	100 h	10 h	1 h	0 h
	SCALE OF TIME			

temperature of 9° C. The average monthly minimum is 11° C so 9° C is unusual without being exceptional. Further south, temperatures in the Congo were reduced and later there was unusually heavy rain at the Kariba dam site, which perhaps indicated the arrival of cold air from the northern hemisphere at south of 10° S.

Weather charts for South America (Fig. 47(a) and (b)) show an outbreak of cold air reaching the equator from the south.

On 17 July the overhead sun is still very near the tropic of Cancer. The chart shows a very slack area of pressure over Amazonia, with light winds. Temperatures at river stations are in the mid-twenties. At Iquitos, where 1200 GMT is 7 a.m. sun time, there is low fog and a temperature of 22° C. The temperature difference between western Brazil 25° C, Manaos 26° C and Belem 27° C is connected with local time differences. The table for Uapes (Fig. 48) shows temperature rises of about 1° C per hour during the period from sunrise to maximum temperature. Notice the temperature at Quito. Along the north coast of South America and in the southern Caribbean the chart shows the easterly trade winds and ships reports suggest a tiny low at 10° N. 52° W. The southern portion of the map shows a frontal depression in eastern Paraguay with rain ahead of the cold front at Belo Horizonte and heavy rain behind in the hills of Parana province, Brazil. In north-east Brazil the south-east trades are blowing.

An area of high pressure built up behind the cold

front in southern Brazil and from the 18th to the 22nd the cold front advanced northwards across the western Amazon lowlands. The chart for 21 July shows it north of the equator lying stationary and becoming weaker on the flanks of the Guiana Highlands. A comparison of the two charts shows a drop of 6° C or more in the colder air mass for stations along the upper Amazon. This is the Amazonia cold spell or *friagem* which may occur from one to three occasions in a year. At Belem ahead of the front conditions have changed little between the two charts. This is true for many stations north of the cold air: notice Quito, Panama, Georgetown, Barbados and Salvador.

On the southern Brazilian coast the cold front has only moved slowly from Rio de Janeiro on the 17th to near Vitoria on the 21st. The ridge of high pressure over the coffee region is giving clear skies with light winds resulting in low temperatures of 5° C at 9 a.m. local time. Such conditions allow air drainage into the valleys and give local frost pockets which affect the position of the coffee plantations. Cold fronts of this type often reach this area in winter and account for most of the cool season rainfall. They are held back at about 15° S. and frontolyse, partly because of the Brazilian High-lands, but also because eastern Brazil, protruding far into the Atlantic, is much influenced by the subtropical high in the southern Atlantic. North-east Brazil, because of this anticyclonic influence, is an anomalous region in this part of the humid tropics.

47(*a*) Synoptic chart, 1200 GMT, 17 July 1957
(*b*) Synoptic chart, 1200 GMT, 21 July 1957

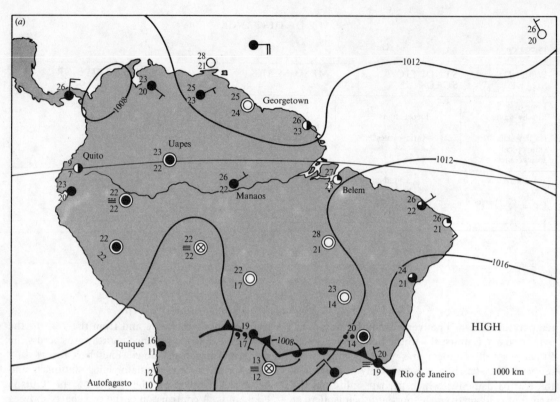

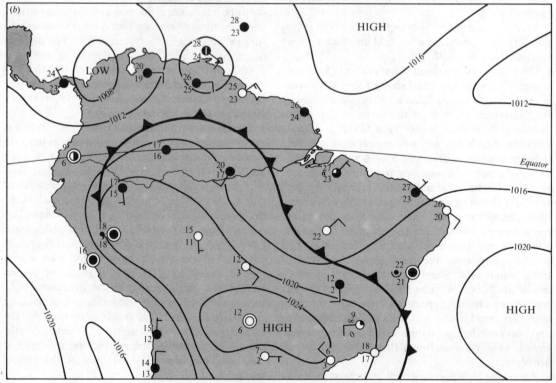

48 Observations for Uapes, Brazil (0° 67°W., height 84 m) for 16–24 July 1957

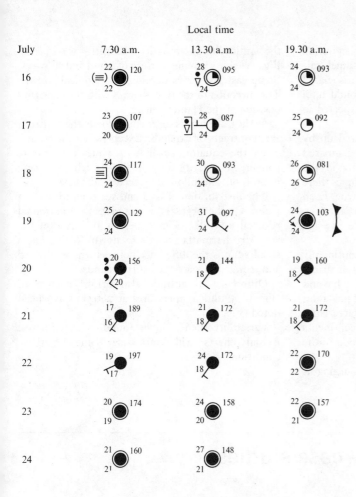

Figure 48 shows a useful series of weather observations for Uapes, on the Rio Negro, almost exactly on the equator in north-western Brazil. They should be examined along with the two charts. Average annual rainfall at Uapes is in excess of 2500 mm. 16, 17 and 18 July are typical equatorial days with cloudy mornings with fog present or nearby, showery afternoons and clearer evenings. Notice the calms or light winds; the daily rise and fall of temperature; the high humidity. There is also a daily rhythm of pressure changes: pressure falls during the morning so a comparison of pressure changes from day to day at the same hour is more useful than a comparison from one observation to the next. Distant lightning on the evening of the 19th heralds the arrival of the cold front and rain and drizzle at the next morning observation with a distinctly lower temperature marks the arrival of the colder air mass. For three days there is a constant light south-west wind with overcast skies and a marked rise of pressure. Only by the 24th are temperatures beginning to get back towards the values of the 16th, showing the gradual modification of the air mass as it stagnates over the equatorial forest region.

Figure 41 for the north Pacific, not far from the equinox, shows the basic surface conditions for the tropical zone in a straightforward way. At, and after, the solstice the equatorial thermal trough moves a greater distance – about 10°. Figure 34 shows the average position of the inter-tropical troughs in the months of greatest displacement. Over the oceans this displacement is generally much less than over the land. Comparison of the Atlantic and Africa illustrates this well. The axis of the subtropical anticyclone also migrates with the apparent movement of the overhead sun but the range of displacement is smaller. Instead the anticyclones tend to become narrower in their north–south, or meridional extent in the summer, and broader in the winter. This accommodates the movement of the thermal trough.

The threefold division is a very useful beginning and by considering the apparent latitudinal migration of the overhead sun, and the associated oscillation of the belts, it is possible to build up a model which depicts the climate to be expected in any latitude.

But this first approach must be followed up by closer individual analyses. Although as compared with mid-latitude regimes the variability in the tropics is lower, statements like 'In the tropics there is no weather only climate' are far from true. Each of the three belts is subject to perturbations – disturbances giving variations from the average conditions (see Table 6.1).

The subtropical anticyclones are continually changing their positions, intensifying and retreating. The eastern ends of the anticyclones are stable and dry but the western ends are moister and unstable: so as they change position, air mass characteristics vary. Only near the western edge of the continents do stable conditions predominate in the roots of the trade winds.

The trade winds change in strength, surging as the anticyclones intensify. Easterly waves (see Fig. 41) give a moving pattern of cloud and showers. Some easterly waves develop into tropical storms. The direction of the trades varies with the changing position of the Hadley cells.

As the trade winds surge or die down, the doldrum area contracts or expands. Except when the trough is on the equator, small lows which concentrate convergent activity form in the trough and bring periods of rain followed by relative dryness.

The distribution of land and sea is an important factor. On the largest scale this gives the monsoonal reversal in India. Slightly less is the scale of the very dry harmattan winds in north Africa. On a small scale is the difference between the windward and leeward sides of even small islands.

Difference in altitude also brings change in climate. Quito temperatures at 2863 m have been noted (Fig. 47).

Variability then is to be found in all climates. Actual charts will emphasise some of these conditions.

7 Tropical climates – case studies

West Africa

The synoptic chart (Fig. 49) for North Africa in mid-July illustrates the main features of the summer weather. As this is just after the solstice, the axis of low pressure, at about 20° N., is very close to the line of the overhead midday sun. The frontal zone separates dry air to the north from moister air to the south. Dewpoints north of the front are very low. In west Africa this very dry air is known as harmattan. It gives clear skies but usually conditions are very hazy due to fine dust in the air. Except along the coast where the sea and especially the cold Canaries Current influences the readings, the temperatures are very high on this midday chart. Several record over 40° C. Temperatures south of the frontal zone are considerably cooler and the dewpoints in the equatorial air are much higher.

Figure 50 shows a cross-section of the atmosphere from north to south along 5° W. The dry harmattan air overlies the moist air from the south. At higher levels the north-easterlies (harmattan) decrease in speed and are replaced by westerlies aloft. Only 600 km south of the surface front is the moist air sufficiently deep for cumulonimbus clouds to develop and give showers. Often these deeper clouds are formed along what are known as 'disturbance lines' which extend from north to south in the moist air. Two disturbance lines are shown on this chart: one between 5° and 10° W., and one about 20° E. giving rain in the north Congo. Thundery rain storms on disturbance lines travel from east to west across west Africa with the upper easterlies which are shown on Fig. 50 overlying the monsoon air. Along the coast more continuous rain often occurs in the on-shore south-westerlies. Rain in sight is reported from a station in the Cameroons. Thus four belts of weather can be distinguished in

49 Synoptic chart, 1200 GMT, 14 July 1967

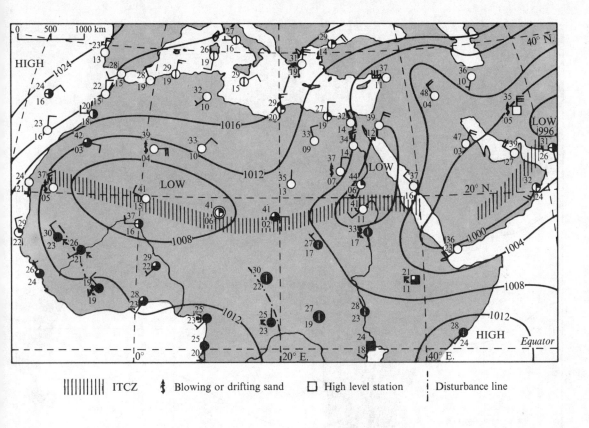

|||||||||| ITCZ   ⚡ Blowing or drifting sand   ▢ High level station   ⋮ Disturbance line

50 Generalised section north to south along 4°W. on Fig. 49

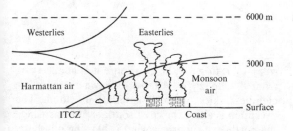

51 Synoptic chart, 0600 GMT, 27 December 1960

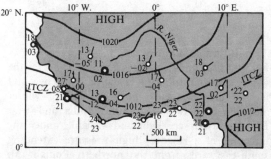

west Africa. From north to south they are: dry, harmattan air; dry but humid air just south of the convergence zone; a region with rain from disturbance lines; and a region of more continuous rain. These four zones move north and south with the overhead sun.

During the winter season the subtropical anticyclone over the Sahara is often well developed and north-east winds on its southern side carry very dry harmattan air far to the south. The inter-tropical

convergence zone reaches the coast though the harmattan season there is very short – usually of one or two weeks' duration. Fig. 51 shows a chart near the winter solstice. Wind directions but not speeds are shown. There is a large difference in dewpoint across the convergence zone. North of the I.T.C.Z. the cloudless though hazy skies allow out-going radiation accounting for the low temperatures on this early morning chart, e.g. 13°C along about 13°N.

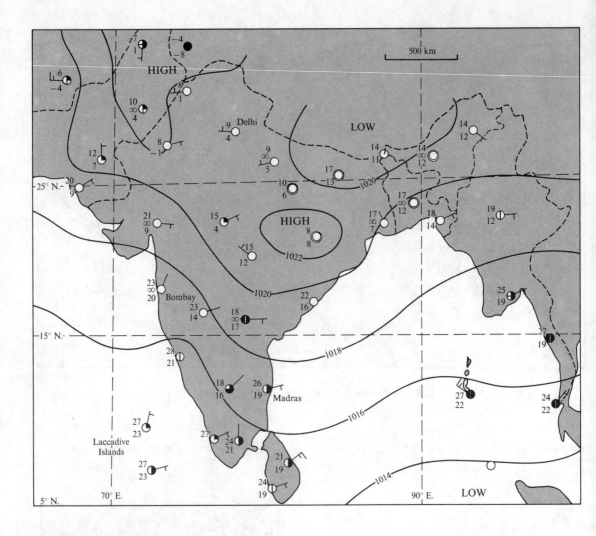

The Indian subcontinent

Figure 52 for mid-January is a typical chart for the winter season in India. Over the Bay of Bengal and its fringing coastlands the north-east or winter monsoon is blowing. In north India and west Pakistan winds are light and variable under the influence of a ridge of high pressure extending south-eastwards from the northern Punjab. An area of low pressure over Iran is giving a gradient for south-easterly winds in the north-west. 0300 hours GMT represents 0700 hours local time in the west and 0900 hours local time along 90° E. The wind speeds and temperatures are representative of morning conditions. Notice that temperatures vary from 24° C at 10° N. to 9° C at 28° N. at Delhi – a general decrease northwards of rather less than 1° C for each degree of latitude. At Madras the average daily range for January is about 10° C,

from 19° C to 29° C; at Cawnpore it is about 15° C, from 8° C to 23° C. Almost everywhere there are cloudless skies or small amounts of cloud. The exception is along the coast south of Madras where the north-east monsoon wind is blowing on-shore. There is no rain shown on the chart though a more complete chart shows there has been rain in the night on the south-east tip of the peninsula. Clear skies are typical of much of India, especially the western Deccan, from November to May.

In winter, north India is affected by low pressure systems. These westerly disturbances often only show a trough at the surface but can be more intense, as Fig. 53 indicates. Between October and April about six perturbations of varying degrees of intensity affect northern India each month but during the whole winter only a few are as intense as

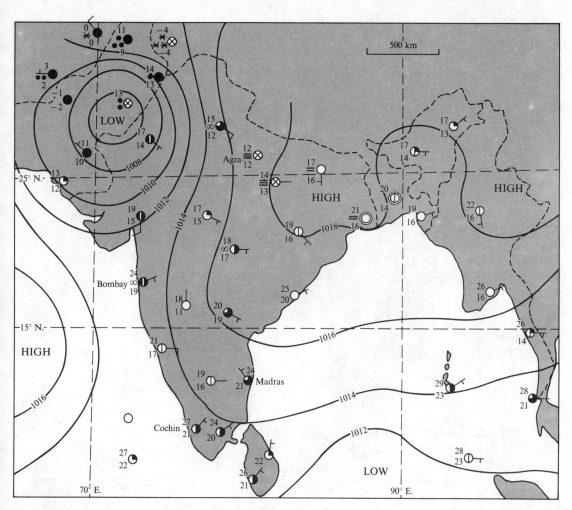

the one on the chart for 26 December. Notice the ridge from the lower Ganges along the eastern Deccan, preserving north-easterlies across the Bay of Bengal, the peninsular tip and Ceylon. The clear skies and the temperature figures accord with those of Fig. 52. The main feature is the depression of 1004 mbar centred over the middle Indus. North and east of the centre is a wide belt of rain. Snow is reported from Srinagar in the mountain belt. Colder drier air, as shown by the dewpoint, has reached Karachi and the lower Indus. Sometimes the colder air sweeps right across India causing a lowering of temperatures in the far south, as the following temperatures for 0700 hours local time from another westerly disturbance show.

	Agra	Bombay	Cochin
5 January	10 °C	15°	28°
6 January	6 °C	13°	28°
7 January	7 °C	12°	27°
8 January	7 °C	12°	25°

The dotted line represents the passage of the cold front.

The way in which these westerly disturbances fit into the general circulation pattern over India will be considered later.

Figure 54 shows a typical synoptic chart at the height of the summer monsoon season. For many locations in the north of India July is the wettest month. Notice the wind directions. Although this is

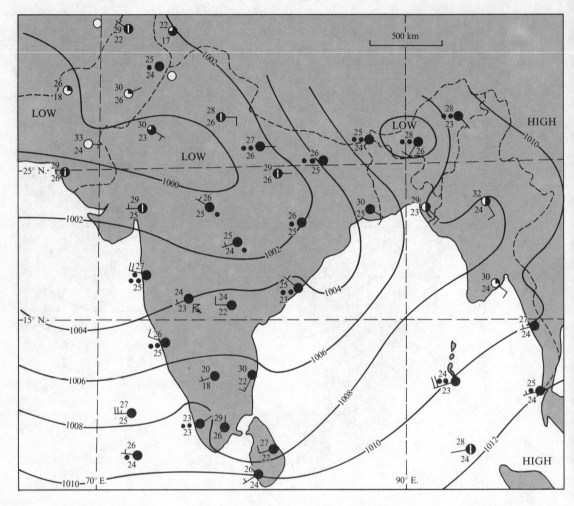

the season of the south-west monsoon the surface winds are by no means all south-westerly. In the Indo-gangetic trough they are south-easterly. Sea level temperatures throughout India show a fair degree of uniformity on this morning chart. Dewpoints are also high indicating a high relative humidity. In mid-July the sun will be overhead at midday near to 20° N. The isobars show a trough of low pressure stretching from Iran in the west across the northern Deccan and then south to 20° N. and away east into central Burma. There is a subsidiary trough down the east coast and a separate low pressure area over Assam. Rain is falling along these trough lines except in the north-west where the air is drier. There is also rain along the west coast.

Figure 55(a) shows an afternoon chart for late August. Notice the very high temperatures in the north-west; the somewhat lower values over the northern Deccan and the lower Ganges and the

lowest temperatures where rain is falling. In the north-west the air is very dry as the low dewpoints show. As in west Africa, the dry air overruns the moist air as shown on the section Fig. 55(b). Only where there is a sufficient thickness of moist air can rainclouds form. Rain is falling in a belt from Delhi to the Rann of Kutch. Nagpur reports a shower in the past weather and Calcutta has had a thunderstorm. There the moist air is deeper. Earlier and later in the summer monsoon season the lows affect southern India.

An elementary explanation of the Indian monsoon compares it with land and sea breezes on a seasonal scale. Central Asia becomes very cold in winter and an intense anticyclone forms there (see chapter 9). Winds blow out to the low pressure over the warm Indian Ocean giving the outblowing north-east monsoon. In summer the situation is reversed and a large thermal depression forms over Asia. Winds

55(*a*) Synoptic chart, 1230 GMT, 25 August
55(*b*) Cross-section from Peshawar towards Calcutta on Fig. 55(*a*)

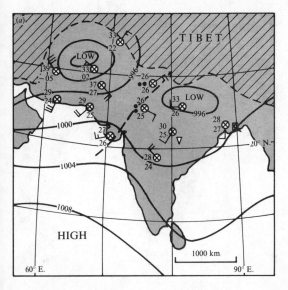

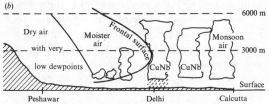

are drawn towards it giving the summer or inblowing monsoon. The Himalayan block is seen as a shield for India from the harsh cold of Asia in winter, and in summer as a barrier deflecting the south-west monsoon up the Ganges valley towards the Punjab. Modern work in climatology suggests a much more important role for the mountain mass north of India. The Tibetan plateau is an enormous block of high ground, 600 km wide in the west and 1000 km wide in the east. It reaches 4 km in height generally, though there are lower valley bottoms and higher peaks. It is a formidable barrier and one of the grand geographical controls on the general circulation.

The macro-geographical pattern in the Indian Ocean segment with warm ocean to the south, high mountains in the centre and the wide plains of the U.S.S.R. to the north give a separate climatic personality to the region.

In winter the Asian continent is intensely cold south to 40° N. and the Tibetan plateau extends this cold region even further south. The winter jet stream is brought to a very southerly position by this southern extension of low temperatures and divides west of the mountain block. One arm flows north and the other arm south of the mountain barrier. It

is the southern limb of the jet stream which is responsible for steering the western disturbances into northern India.

In spring as the overhead sun moves north the extensive snow cover of the Himalayas and Asia, with its high albedo, is slow to melt, but north India, protected by the mountain wall, warms more quickly than would be expected. So the thermal contrast between cold and warm remains far south along the Himalayan line and the jet stream continues to blow far south. Only in early April does the snow line begin to recede rapidly; the Asian continent is also now warming quickly and the thermal contrast weakens. Once the snow line has receded the mountain block assumes a new role. As explained in chapter 1, the atmosphere is heated from below: the sun's rays heat the ground which then heats the atmosphere. The expanse of high land north of India is heated in this way. It then heats the air above it, so the air above Tibet at a height of 4 km, say, is almost as warm as much lower air over India to the south and therefore much warmer than air at the same level to the south. Warm air rises. Therefore over north India at intermediate levels in the troposphere is rising air. This draws in currents from the south and helps to spread the south-west monsoon quickly over the subcontinent. It spreads from south to north in about three weeks – the burst of the monsoon.

The zonal land–sea distribution has an important effect too. By mid-summer the zone of most intense heating is around 20–25° N. Uniform sea, as in the Pacific, gives little movement of the pressure belts. Uniform land area, as in Africa, gives somewhat greater movement. In the Indian Ocean, the land–sea distribution, with sea which heats up slowly in the south, and land which heats up quickly further north, is ideal for the greatest displacement of the pressure belts. So the equatorial trough transfers to the north of India.

Figure 56(*a*) shows that the so-called burst of the monsoon occurs in a few weeks. Between 1 and 15 June the south-west monsoon advances over almost the whole of the Deccan, Bangladesh and Assam. While the jet stream is south of the Himalayas, advances far north cannot occur. But the jet stream cannot retreat slowly – it must either be south of the mountains or not there at all – so the process is not gradual but happens quickly once the thermal conditions are satisfied. The change from one regime to another is sharp.

In describing the climate of India it is usual to divide the year into: (*a*) the cool season lasting to the

41

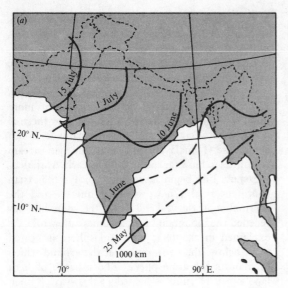

end of February; (b) the hot season from March to May; (c) the wet season from June through the summer. The change-over from wet season to cool season depends upon the latitude. In north India October can be regarded as the change-over; in the south, it is mid-December (see Fig. 56(b)).

Except in the north, winter is practically rainless and temperatures average 20–30° C with a good daily range. The hot season starting in March is still rainless but day temperatures begin to soar to 35° or 40° C. In May the north winds are replaced by southerly humid air which at first brings no rain. (It is not sufficiently deep, cf. west Africa.) Rangoon and Calcutta have some lighter rains in May before the main monsoon reaches Bombay.

At last the branch of the jet south of the Himalayas weakens and the monsoon bursts. The south-west monsoon current then prevails throughout India, except in the north where it is deflected by the Himalayan mass and attracted by the monsoon trough to blow from the south-east. The monsoon trough is the Indian name for the tropical convergence zone which is so far displaced here that it is better called by this alternative name. The south-west monsoon current is $6\frac{1}{2}$ km deep in southern India but $1\frac{1}{2}$ km shallower over the Gangetic plain. Above this current is a layer of easterly winds.

The monsoon current sweeping across the Arabian Sea becomes effective as a rain producer when it reaches the western Ghats. The lee of the Ghats is much drier. Apart from the mountain areas the other heavy summer rainfall region is the monsoon trough. During the summer monsoon season there are two main patterns: active monsoon

as illustrated by Fig. 54 and the break in the monsoon or weak monsoon. During the active monsoon, depressions move from the Bay of Bengal into northern or central India, but even without active depressions periods of continuous rain are widespread. The origin of these monsoon depressions is not certain. Some meteorologists have connected them with waves in the upper easterlies.

Breaks in the monsoon can last from three to ten days. They are few and short in July but longer and more frequent in August and September. A typical chart of a break in the monsoon shows a weak pressure gradient over peninsular India. Rain over the western Ghats is reduced; central India is dry. The monsoon trough moves to its most northerly position and rainfall is normal or above average along and near the foot of the Himalayan mountain wall. In the extreme south over Cochin and Ceylon shallow surface lows form. The convergence appears to re-form here and move north again to bring a renewal of active monsoon.

If the monsoon is late, and if weak monsoon predominates throughout a summer, the rainfall received may be reduced disastrously and the monsoon can be said to have failed.

In late September and early October the circulation pattern near the Himalayas returns to its winter regime and the monsoon begins to withdraw. At this period over the Bay of Bengal and the Arabian Sea some depressions develop to hurricane strength and bring occasional heavy rain and strong winds. In particular strong southerlies blowing over the channels of the Ganges and Brahmaputra can bring devastating floods as happened in 1970 and

57 Synoptic chart showing four hurricanes,
18 September 1968 (central pressures of hurricane
given in brackets)

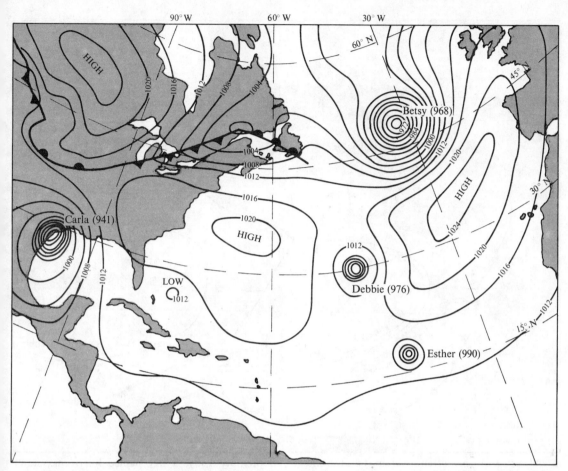

1971. Fig. 56(*b*) shows that the withdrawal of the monsoon is a much slower process over southern India, and that Madras gets its main rainfall at this season of the retreating monsoon.

Tropical cyclones

Every year in the tropics violent winds associated with a few intense areas of low pressure do immense damage. These storms have special local names in different areas (see Table 7.2, p. 46). However, not all tropical storms reach disastrous proportions. In the West Indian region the nomenclature used is as follows:

Tropical depression   few closed isobars, winds below 64 km/h

Tropical storm   winds between 64 and 116 km/h

Hurricane   winds above 116 km/h

Fig. 57 shows a very unusual chart for September 1961 with four fully developed hurricanes in the Atlantic and Caribbean at the same time. Notice the circular, concentric isobars and the steep gradient.

The diameter of a hurricane varies, usually from 80 km to 320 km, but some are even larger. The average number of tropical storms in the North Atlantic in the season is 10. Table 7.1 shows the numbers in recent years.

Year	Total number of tropical storms	Storms of hurricane intensity
1965	6	4
1966	11	7
1967	8	6
1968	7	4
1969	13	10
1970	7	3
1971	12	5

Table 7.1   *Tropical storms in the North Atlantic, 1965–71*

Forecasters choose an alphabetical series of girls' names before the season begins and each storm is then referred to individually by its name. This is

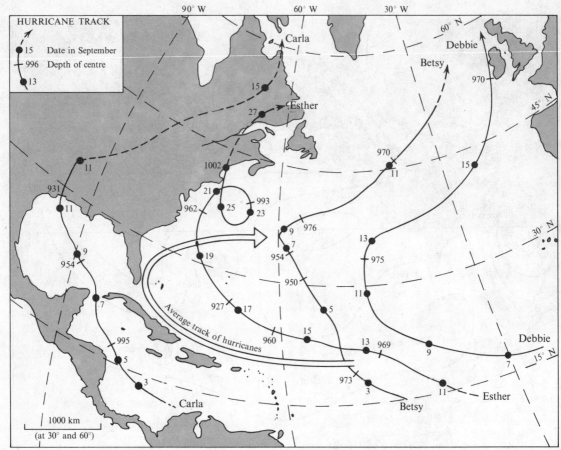

HURRICANE TRACK

15 Date in September

996 Depth of centre

13

Carla

Debbie

Betsy

970

60° N

45° N

Esther

15

27

1002

970

11

15

11

21

993

962 25 23

931

11

9 976

7

13

11 954 975

9

954

19

950

30° N

7

11

927 17

5

995

15

13 969

5 960

13 Debbie

3 973 9 15° N

Carla 3 11 Esther

Betsy 7

1000 km

(at 30° and 60°)

Average track of hurricanes

90° W 60° W 30° W

useful since storms often overlap because they remain on the chart for a number of days, e.g. on Fig. 57, Betsy, 2–11 September; Carla, 3–15 September; Debbie, 7–16 September; Esther, 11–26 September. The wind speed in hurricanes commonly exceeds 120 km/h but exact maximum speeds are hard to measure because of the destruction of recording apparatus. The whole tropical cyclone moves bodily at more moderate speeds of 15–50 km/h along its track. Sometimes cyclones become stationary; sometimes they loop and follow complicated tracks. The tracks for Betsy, Carla, Debbie and Esther are shown on Fig. 58 along with the average track. If all known cyclone tracks were plotted no area in the Caribbean and west Atlantic would be free, but the most common track is parabolic, guided by the subtropical anticyclone. Debbie had a rather extreme easterly track and passed western Ireland and reached Iceland with hurricane-force winds. This storm was responsible for 17 deaths in Ireland. Carla had a southerly track not uncommon for early and late storms.

Figure 59 is a satellite photograph taken on 14

September 1967. Three named hurricanes stand out clearly over the Atlantic; a fourth hurricane can be seen over the Pacific south of the peninsula of lower California (17° N. 108° W.). The cloud over Texas and the central Mississippi valley is connected with a warm frontal trough. Eastern U.S.A. is under the influence of an anticyclone. The less dense cloud near Lake Michigan is fog or low stratus. It can be compared with the stratus over the cold current along the Californian coast and contrasted with convective cloud in the hurricanes. The sizes of the hurricanes can be estimated from the known length of 10° of longitude.

In the centre of the tropical cyclone is a region sometimes clear, sometimes cloudy, with light winds or calms. It shows up on the satellite photograph (Fig. 59) of Chloe (31° N. 55° W.). This is the eye. Its diameter can be estimated. Around the eye are hurricane winds, heavy rain and thick cumulonimbus clouds caused by the quickly ascending air. A well developed cyclone may give 500 mm of rain in a day, but because it is travelling this does not fall in one place; though 250 mm a day in one

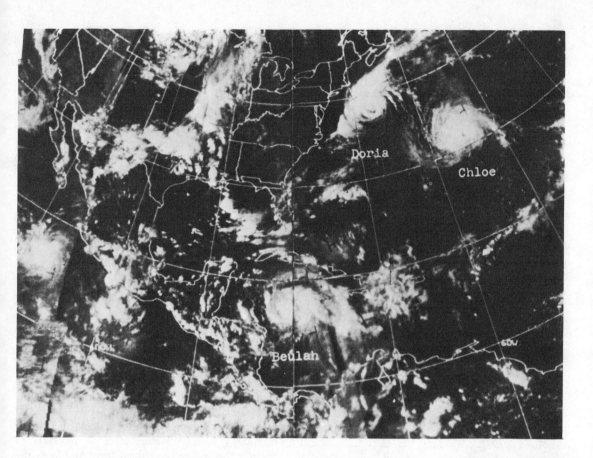

locality is often exceeded. Hurricanes are immensely destructive. At sea the waves are huge and the sea confused. On land the force of the wind does great damage. Rain may cause flooding. Shores suffer from waves and swell. Hurricane Betsy, the only hurricane in 1965 to make a land-fall, can be cited as an example. It was the most destructive on record. First located in mid-Atlantic on 23 August, Betsy crossed the Lesser Antilles on the 27th and deepened to 942 mbar by 2 September. She appeared to be passing away from the mainland but on 5 September her track changed from its parabolic curve and she began to move south-westwards. The eye passed just north of Nassau in the Bahamas causing one casualty and £5 million damage, and reached the Florida Keys by 7 September. Rainfall at the Keys was 300 mm and four people lost their lives. During the 8th Betsy crossed the Gulf of Mexico and went ashore near the mouth of the Mississippi on the 9th. Here the eye was 64 km across. Much damage was caused by high water along the Gulf coast and in the Mississippi mouth. Hundreds of barges were sunk between New Orleans and Baton Rouge. The river

at New Orleans rose 3 m during the hurricane and there were 58 casualties. Betsy moved north and by 11 September reached the depression stage. In Arkansas there was damage to the cotton and rice crops but this was offset by the increased soya bean yield due to the much needed rain of between 76 and 178 mm. There was much damage to cotton and pecan nut crops along the Gulf. In southern Florida 25–50% of the citrus fruit was blown from the trees and in some counties 90% of the avocadoes were lost. There were also losses in oil equipment off the Louisiana coast.

The origin of tropical cyclones is still debated. Some meteorologists have emphasised the necessity for some out-flow mechanism in the upper troposphere which allows pressure to fall at the surface and low level in-flow or convergence to begin. If an easterly wave in the trade wind zone (see Fig. 41) coincides with a high level cold trough in the westerlies, convection through a deep layer can start and the necessary fall of pressure can begin. Many Caribbean hurricanes have now been traced back across the Atlantic and shown to have con-

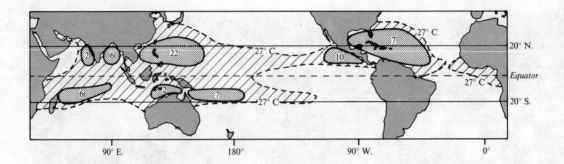

nections with easterly waves leaving the west African coast. In 1971, of 56 African systems passing Dakar 38 were tracked through the Caribbean into the Pacific.

Other meteorologists have tackled the problem of origin using a more geographical approach and have suggested the development of a tropical depression is much more a matter of the chance occurrence of a number of factors simultaneously. They begin by pointing to the relative infrequency of tropical cyclones. The world average is about 60 per year. Fig. 60 and Table 7.2 show the average numbers in each area. As more satellite weather information becomes available, giving more complete ocean coverage, these average figures may need some adjustment upwards.

Location	Local name	No.	%
North Atlantic –			
Caribbean and Gulf of Mexico		7	11
North Pacific – off W. Mexico		10	16
North Pacific – East Indies	typhoons		
Philippines	baguios	22	36
North Indian Ocean –			
Bay of Bengal		6	10
Arabian Sea		2	3
South Indian Ocean		6	10
Off N.W. Australia	willy-willies	2	3
South Pacific Ocean (including			
off N.E. Australia)		7	11
		62	100

Table 7.2 *Average number of tropical cyclones per year*

As tropical storms have an average life of twelve days any world synoptic chart could be expected to show two, if they were evenly distributed throughout the year. But they are not. As Table 7.2 shows, about 75% occur in the northern hemisphere, where they occur mainly between June and October inclusively.

Extra-tropical depressions average about 40 each week. In the U.S.A. alone some 200 tornadoes are reported each year. At any time there are thousands of thunderstorms somewhere on the earth. This demonstrates that tropical cyclones are really rather uncommon but because of their energy and possible destructiveness they have an importance despite their comparative rarity.

The main energy source for a tropical cyclone is the latent heat released by condensation from warm moist air. Fig. 60 shows that they form over the sea with a minimum surface temperature of 27°C. Tropical storms do not develop very close to the equator because the deflecting force is too small: winds quickly blow in towards low pressure and so large pressure gradients cannot exist. With a strong rotation a large deflecting force is necessary to balance the strong centrifugal force and keep the winds parallel to the isobars. In this way low pressure is preserved. Tropical storms are not found within 5° of the equator for this reason. When tropical storms go overland they usually decrease considerably in strength because their source of energy from warm moist air is cut off. Friction causes air to cross the isobars and the low pressure begins to fill up and become less intense.

Cyclones form on the poleward side of the inter-tropical convergence. Here it is necessary to distinguish two forms of inter-tropical convergence. Where trade winds from each hemisphere run together the convergence has no belt of calms and can be labelled a trade-wind inter-tropical convergence. When a belt of calms occurs along the convergence this is a doldrum inter-tropical convergence. Just poleward of the doldrum trough tropical storms can develop in an air mass which is deep and homogeneous and shows little or no change of wind with height.

In these regions convergence caused by a backing of the wind (about 10°) between cloud base and the sea surface is sufficient to start uplift. Latent heat of condensation adds heat to the rising column of air. Homogeneity of the air mass ensures that large cumulus can penetrate high into the troposphere. This mechanism suggests that tropical storms develop from very large cumulus clouds. What is required is that the cumulus column shall become warmed to a temperature of about 1° c greater than its environment. Such a degree of warming will give a pressure fall at the surface of about 4 mbar, which is sufficient for storm formation to begin. Evidently it is not easy for this minimum 1° c temperature rise to be attained. Differential absorption of radiation from the sun's rays by the cloudy column as against the clear circumambient atmosphere may help, as does the release of latent heat of condensation inside the cloud. Once the initial heating necessary to give the pressure fall has taken place intensification will follow. Air converging on the low pressure area will be uplifted and energy for the storm will be provided by the latent heat set free on condensation. This mechanism suggests the important criterion is how to keep heat in the column of air at the beginning of the process: the theory requiring an easterly perturbation at the surface and a cold trough aloft is a mechanism for losing heat in the upper troposphere.

Figure 60 shows the regions of origin of tropical cyclones. A band of 5° of latitude either side of the equator is clear. Beyond the 5° line cyclones form in all oceans except the South Atlantic which is the area where the inter-tropical convergence never penetrates far south of the equator. Cyclones occur at the season of high sun when the inter-tropical convergence moves from its equatorial position. Early in the season cyclones form nearest to the equator: as the season advances they form in higher latitudes and in the late season again nearer the equator.

8 Mid-latitudes – west coastal areas

The importance of the relationship between the development of ridges and troughs in the upper westerlies and surface anticyclones and depressions was explained in chapter 3.

The weather experienced in a particular sector of the westerly air flow at any time of the year is largely dependent on the amplitude of the Rossby waves. If the amplitude is small the flow in the upper westerlies is latitudinal, i.e. west to east, as is the axis of the jet stream which steers the frontal depressions. When the amplitude of the waves is large there is a considerable transport of air from tropical regions polewards on the western side of the upper ridges and from polar regions equatorwards on their eastern sides, i.e. the air flow is meridional as are the tracks of the jet stream and frontal depressions.

The changes in the amplitude of the Rossby waves are consequently linked with the changeable character of the weather in mid-latitudes which is particularly characteristic of the eastern Atlantic and western Europe.

Western Europe

Summer

Figure 61 illustrates the weather experienced over the eastern North Atlantic and Europe when the upper westerly air flow was almost latitudinal. A family of polar-frontal depressions, which were travelling at *c.* 40 km/h, was giving a spell of unsettled weather in those parts of Europe between *c.* 45° N. and 60° N. as they penetrated deeply into the continent. As is common when depressions are relatively shallow and travelling quickly, the frontal rainbelts were narrow and rainfall light. Cool cloudy weather, with light showers, was occurring in the moist mP air in the rear of the depressions. In the pronounced ridge of high pressure over north-west Russia, however, the temperatures were appreciably higher as a result of the heating of the land during the long hours of daylight.

Anticyclonic conditions dominated the weather in the mT air mass south of the frontal zone and temperatures were up to *c.* 15° c higher than to the

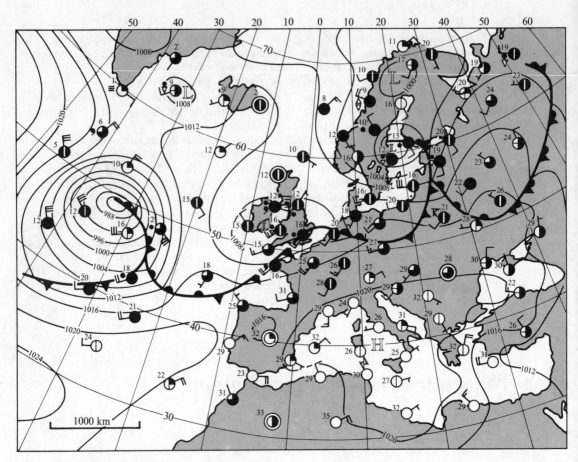

north of it. The Mediterranean area was experiencing the fine weather typical of the time of the year but more cloudy conditions occurred over France and central Europe in the humid air from the Atlantic.

The persistence of a west to east air flow and the frequent passage of depressions are the causes of cool, wet summers in western Europe. Such weather was also experienced by the British Isles and adjacent parts of Europe in the early 1960s and in May, June and July 1972. The summer of 1965 was particularly cool and wet and Fig. 62, based on the recordings made at Sheffield, illustrates the very variable weather conditions associated with the frequent passage of depressions following tracks well to the south in June and July. The daily maximum and minimum temperatures were slightly above the long-term averages in June but below them in July. Sunshine totals were below average in every month, especially in July. The total rainfall for the period was 136 mm, as opposed to the 1901–50 average of 114 mm. For the summer as a whole rainfall was below average in July and

August, the wet months being May, June and especially September when the amount was 315% of the average.

During the five months 29 depressions crossed the British Isles into neighbouring parts of Europe and high pressure conditions occurred in Britain on only 20% of the days.

When western Europe has an unsettled, cool, wet summer resulting from frequent incursions of mP air, eastern Europe has much more settled weather with above average temperatures resulting from the poleward transportation of warm air on the western flanks of the blocking anticyclones which frequently dominate the area.

Dry summers in western Europe occur when blocking anticyclones form and persist in the vicinity of the British Isles. The period May to September 1959 gave the driest summer in England and Wales for 200 years and the dry spell persisted into mid-October. The total rainfall at Sheffield for May–September was only 115 mm. September in particular was exceptional with only 5 mm of rain, as opposed to the average of 53 mm (Fig 63).

62 Daily weather summary for Sheffield, England, June and July 1965

63 Daily weather summary for Sheffield, England, June and July 1959

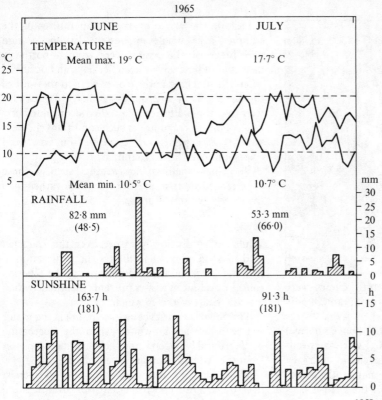

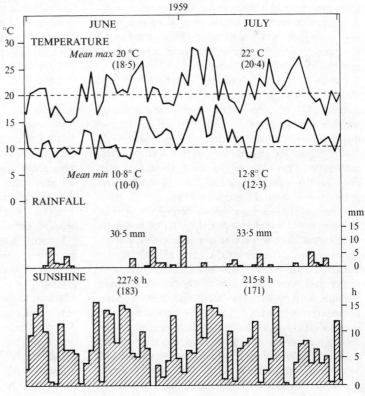

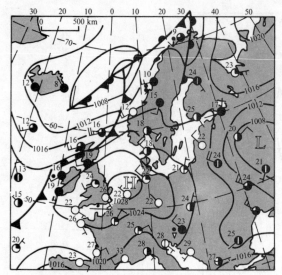

In May, when the blocking anticyclones were mainly centred to the north of Britain, rainfall only occurred on three occasions when weak fronts crossed the British Isles. The above normal temperatures experienced (Fig. 63) were the result of day-time heating under anticyclonic clear sky conditions. In June and July the main anticyclones were centred to the south-west of the British Isles with pronounced ridges extending across north-west Europe. During these months there were only two occasions when the anticyclonic conditions were interrupted by marked cyclonic activity and appreciable rainfall was experienced.

Figure 64 shows the synoptic situation on 7 July when a blocking anticyclone centred over the southern North Sea dominated the European weather west of *c*. 25° E. The weather was fair or fine except along the Atlantic margins where weak frontal waves and depressions were tracking north-eastwards. The SW.–NE. frontal alignment was common throughout June and July, as was the low pressure over central European Russia.

Whereas the summer in western Europe was warmer and drier than normal, parts of northern and central Europe were cooler and wetter, due to incursions of arctic air on the eastern flanks of the blocking anticyclones. Southern Europe was also cooler and wetter because of the above average cyclonic activity over the Mediterranean. In June above average rainfall fell in a belt extending from north-west Spain to south-east Europe and severe flooding occurred in Austria and Bavaria on one occasion. Severe thunderstorms were experienced in central and southern Italy and some Spanish stations received four times the average rainfall.

Despite the above average temperatures of the summer as a whole in western Europe a high percentage of the winds were from a northerly direction. These gave the cooler days and occurred when the blocking anticyclones were to the west of the British Isles and quasi-stationary depressions were over the Baltic Sea area. When strong northerly flows of mP air occur under such conditions showery weather is experienced, especially on windward coasts and over high ground.

Very long spells of dry weather such as those experienced in 1959 are the exception rather than the rule in western Europe.

Winter
Whilst generally low pressure over the Eurasian land mass in summer facilitates the penetration of Atlantic depressions, the dominance of the Siberian cold anticyclone in winter prevents their penetration into the heart of the continent.

The Siberian anticyclone (see p. 66) is the major source of cP air in winter when the fluctuating battleground between it and the maritime air masses from the Atlantic is mainly west of 20° E.

When blocking anticyclones develop and persist over eastern Europe the circulation over western Europe tends to be dominantly zonal and the winter weather mild and unsettled because of the dominance of maritime air masses and the frequent passage of deep frontal depressions.

Such conditions were characteristic of the winter of 1966–7 and Fig. 65, which shows the daily temperatures and rainfall recorded in Sheffield between 14 December 1966 and 10 March 1967, illustrates the general character of the period on the Atlantic coastlands. During this period of 87 days the daily maximum temperature was above 0° C throughout and the daily minimum temperature fell below this level on only 13 occasions. Measurable rainfall fell on 49 of the days and there were three long wet spells associated with the passage of depressions.

Figure 66 shows the synoptic chart at the beginning of one of the wet spells. On this occasion an anticyclone was dominant east of 20° E. and the southerly winds on its western flank were giving relatively high temperatures for the time of the year in European Russia. Over the north-east Atlantic and western Europe a large, very deep depression was centred at 57° N. 16° W. In the steep pressure gradient between the centre of the depression and the anticyclone centred near the Strait of Gibraltar westerly winds reached speeds

65 Daily weather summary for Sheffield, England, December 1966 to March 1967, showing (a) daily maximum temperature, (b) daily minimum temperature, and (c) daily rainfall

66 (*below*) Synoptic chart, 1200 GMT, 19 February 1967

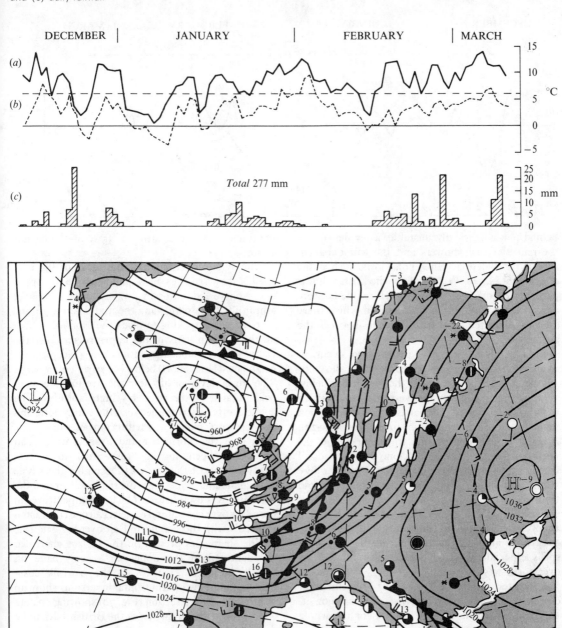

of up to 110 km/h in the mild rmP air north of the cold front. The 500 mbar chart also had a steep pressure gradient over the Atlantic between 40° N. and 50° N. which picked out the strong jet stream. Over France the jet stream divided: the northern branch steered the main frontal system north-

eastwards towards the Baltic whilst the southern branch steered the Mediterranean depressions eastwards.

The rapidly occluding frontal system of the Atlantic depression travelled quickly across western Europe in the vigorous circulation. The rmP air

67 Daily weather summary for Sheffield, England, December 1962 to March 1963, showing (a) daily maximum temperature, (b) daily minimum temperature, and (c) daily rainfall

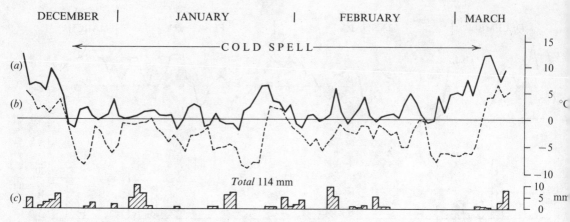

behind it was markedly unstable, as evidenced by the rain and hail showers and the thunderstorm reported at Valentia. When polar lows, such as the one at 50° N. 40° W., develop the instability is particularly marked. The frontal rainbelts were narrow, as is common in situations in which there are only slight temperature differences between the air masses.

Mild spells of winter weather also occur in western Europe when a warm anticyclone develops and persists in the vicinity of the North Sea below an upper warm ridge on the western flank of which mT air is transported polewards. In such a situation night-time radiation fog often forms (see Fig. 22). Atlantic frontal depressions travelling round the northern flanks of the anticyclone give periods of frontal snow in northern and eastern Europe where strong outbursts of arctic air occur behind the cold fronts.

Periodically the cP air from the cold Siberian anticyclone extends its influence to the Atlantic coastlands and dominates the weather for long spells, especially when a blocking anticyclone forms over north-west Europe.

A winter which will be long remembered, especially in the British Isles, was that of 1962–3. In Great Britain it was the coldest for the whole country since 1829–30.

Figure 67 shows the temperature conditions recorded in Sheffield between 14 December 1962 and 10 March 1963, the actual cold spell lasting from 23 December to the 6 March. During the cold spell night minimum temperatures above 0° C were only recorded on three days at the end of January and day-time maxima only exceeded 5° C on nine occasions. Winds between south and west were recorded on very few occasions during the cold spell, whilst those between north-north-east and south-south-east, i.e. from the direction of the

continent, were recorded on 35% of the observations.

The synoptic situation at the onset of the cold spell is shown in Figs. 68(a) and 68(b). In the upper troposphere the Rossby waves were of large amplitude (Fig. 68(b)) and the mid Atlantic frontal systems were tracking towards the Norwegian Sea on the western flank of the upper ridge. Over the Baltic Sea a blocking anticyclone had become established (Fig. 68(a)) below the warm upper ridge. The lower layer of air near the ground was very cold because of the relatively high latitudes. The cold easterly airstream between the anticyclone and low pressure over the Mediterranean caused the front over the North Sea and France, which had previously been moving eastwards, to retreat westwards and by 0600 on 23 December the whole of England and Wales was covered by cP air and the cold spell was established. The low temperatures in the cP air over central and eastern Europe contrasted markedly with those in the mT air over the Atlantic.

The air masses present on the surface chart can be identified in the upper air ascent made on 23 December at Hemsby on the Norfolk coast (Fig. 69). The two warm frontal inversions show up clearly and allow a section (Fig. 70) showing change of air mass with height across the British Isles to be drawn.

The short track across the warmer waters of the North Sea from Holland to Norfolk had resulted in warming at the surface to give an almost D.A.L.R. in the cP air below 940 mbar. Above the inversion was warmer maritime air from the Atlantic with still warmer mT air above the near isothermal layer from 700 mbar to 650 mbar.

During the cold spell the blocking anticyclone did not remain stationary. At times it lay to the west of the British Isles; at other times it extended over the British Isles. Throughout the cold period the

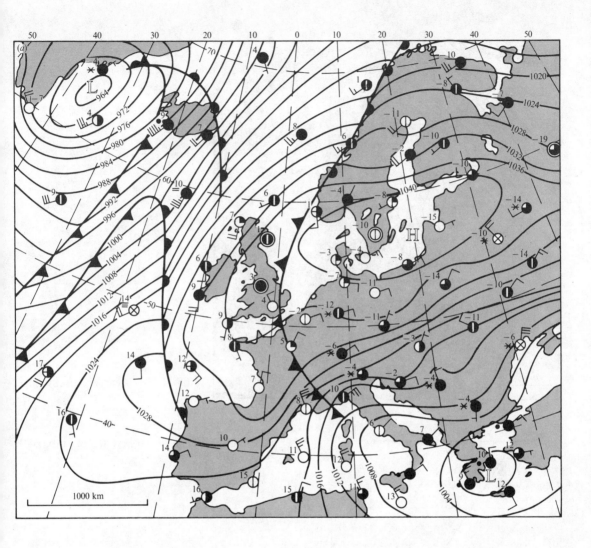

weather in Britain alternated between spells when easterly winds gave much cloud, low maximum temperatures and moderate snowfalls and spells of fine or fair weather with little precipitation and low minimum temperatures in association with anti-cyclonic conditions.

Figure 71 shows temperature anomalies for most of the northern hemisphere for the 'cold' winter of 1962–3. England and Wales were 4° C colder than normal, most of Germany 6° C colder. The Russian arctic was also unusually cold. Most of the U.S.A., except for the western Cordillera, had below average temperatures, with much of the Mississippi basin 4° C colder than normal. Areas with a positive

anomaly were the Canadian arctic, Siberia and the southern U.S.S.R. and the Middle East.

The main point brought out by the map is that temperatures below normal in one region are accompanied by above normal temperatures elsewhere and for the whole of the hemisphere a balance is more or less established. If the areas with anomalies of + and −4° C are traced on to squared paper the reader can convince himself of this. Over the whole season the hemisphere has balanced its heat budget (see chapter 3). A general eastward displacement of the four great centres of action in the northern hemisphere produces anomalies of this type.

53

68(*b*) 500 mbar chart, 0000 GMT, 23 December 1962
(heights in metres)

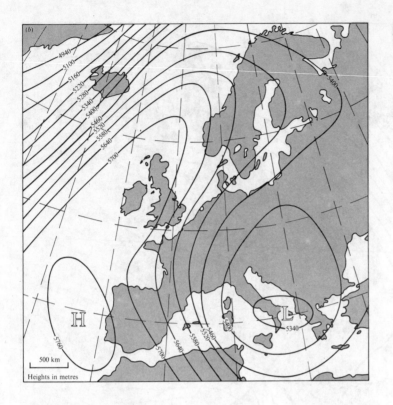

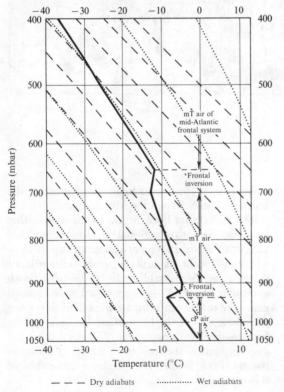

69 (*left*) Upper air diagram, Hemsby, Norfolk, 1200 GMT,
23 December 1962

- - - Dry adiabats   .... Wet adiabats

70 Section across synoptic chart, Fig. 68(*a*)

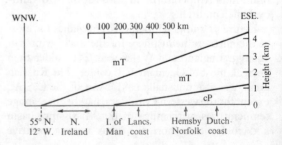

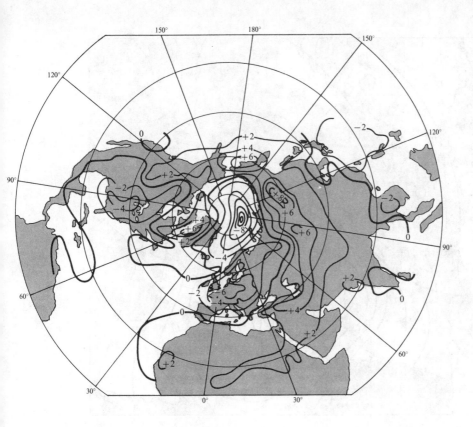

Mediterranean region

As is well known the Mediterranean lands normally experience higher temperatures, greater amounts of sunshine and fewer rain days than western and central Europe. During winter unsettled weather occurs frequently but the summers have a high frequency of calm weather.

Summer

During the northern summer the Azores anticyclone intensifies and becomes the dominant pressure system over the Atlantic between *c.* 20° and 45° N. As a ridge from this develops towards the Alps the summer weather in the Mediterranean becomes mainly calm and sunny from about mid-June to about the end of September (see Figs. 61 and 66). Periodically, however, as in 1955 and 1959, the Azores ridge extends towards Britain instead of towards the Alps (see Fig. 64). When this happens arctic air on the eastern side of the ridge may penetrate into the Mediterranean via the Alps, Balkans or Black Sea. This gives disturbed weather with marked instability along the cold fronts

heralding the arrival of the cold air itself. Such a situation is however rare. Even more uncommon at this season are cold pools in the upper air over the Mediterranean (pp. 11–12) but if they do form, intense thunderstorms result.

When summer Atlantic depressions are moving across Europe cold fronts periodically penetrate into the Mediterranean. They are, however, usually weak and the mP air behind them stable, since subsidence takes place in the ridge of high pressure which normally occurs behind the front. Nevertheless the mP air is often convectively unstable in the upper layers and showers and thunderstorms occur over mountain ranges such as the Apennines due to the orographic uplift.

Particularly heavy thunderstorms occur over the Lombardy Plain and the north Adriatic during the summer months following the formation of thermal lows. They also occur when unstable mP air crosses the Alps and overruns warm, moist unstable air. As a result the Lombardy Plain and the north Adriatic do not experience the summer drought of much of the Mediterranean.

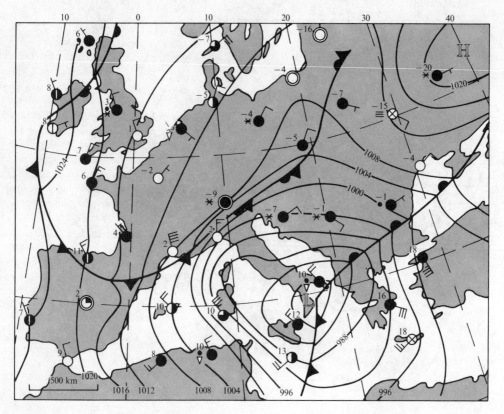

Winter

The onset of the variable weather occurs about mid October when falls of pressure take place over the Mediterranean ranging from 10 mbar in the west to 3–4 mbar in the east and south-east. This follows the first penetration of deep, cold air behind a cold front.

Invasions of deep mP and more rarely mA air occur mainly through the Rhône valley, Strait of Gibraltar and the Carcassonne gap. The average number of such incursions is 5 or 6 a month during the winter. Since the Mediterranean Sea surface is warm (February temperatures range from 8° C in the north Adriatic and 12° C in the north-west to 17° C in the south-east) the cold air is warmed and its moisture content increased during its passage across the Mediterranean. Great instability results and the large cumulonimbus clouds which form, with tops often between 6½ and 13 km, give heavy showers and thunderstorms. From eastern Europe outbursts of cP and cA air also take place.

The interaction between these cold air masses and 'Mediterranean' air (polar air which has stagnated and increased in warmth and moisture content and is convectively unstable) follows cyclogenesis, especially in the form of lee depressions over the

Gulf of Genoa. As they develop, cT air from the Sahara is drawn into their circulation and frontogenesis takes place.

Figure 72 is an example of a frontal depression which originated in the Gulf of Genoa and then moved south-eastwards and became centred over southern Italy at 1200 GMT on 12 January 1960. The temperatures in the cT air of the warm sector were about 6° c higher than those in the showery mP air stream behind the cold front. Ahead of the warm front much colder cP air over south-east Europe had been drawn into the circulation.

As the depression deepened the arctic air over the more northerly parts of Europe, and the arctic front which marked its leading edge, were drawn into its circulation. At 1200 GMT the cA air and the arctic front ahead of it had just entered the north-western Mediterranean. In the north-western part of the depression the pressure gradient was particularly steep, due to the diversionary influence of the Alps on the airflow. The Alps were also retarding the southward movement of the arctic front. A strong wind of c. 75 km/h was blowing at the mouth of the Rhône valley where the sky was cloudless. This strong cold wind was the *mistral*,

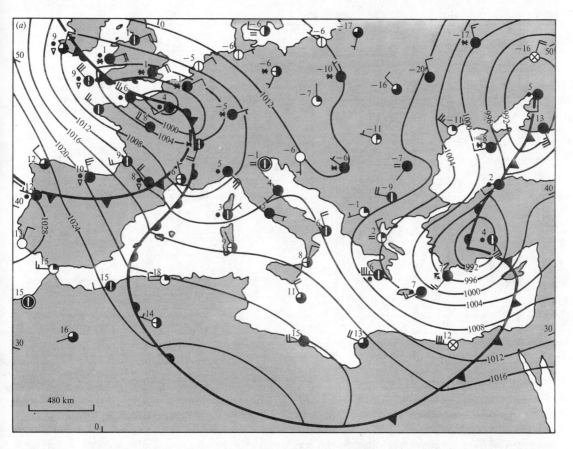

one of the local ravine winds which occur in mountain valleys or gaps around the Mediterranean. When an airstream is flowing along such valleys the winds strength increases as a result of the funnelling and constricting effects of the valley sides.

A strong mistral occurs in the north-western Mediterranean on an average of 103 days per year. It blows for periods ranging from a few hours to 12 days but most frequently persists for 3 days. The average seasonal occurrence is:

Dec.-Feb.	Mar.-May	June-Aug.	Sept.-Nov.
29 days	32 days	25 days	17 days

Similar winds are the *Bora* of the north Adriatic, the easterly Levanter (see Fig. 61) and the westerly Vendale (Fig. 73(*a*)) of the Strait of Gibraltar.

Fig. 73 shows an occasion when the Mediterranean Sea and its borderlands were experiencing a variety of weather conditions. The depression centred over the north-east Black Sea had travelled across the Balkans from northern Italy in the preceding 48 hours. In its rear an arctic air stream crossed western and central Europe into north

Africa. The upper air ascent for Malta (Fig. 73) shows that the arctic air over the island reached 800 mbar. Up to 900 mbar the lapse rate approximated to the D.A.L.R. and then to the S.A.L.R. up to the 800 mbar level. Between 800 mbar and 725 mbar the ascent passed through the frontal mixing zone inversion into the tropical air aloft. This inversion formed the ceiling for cumulus cloud tops over the Mediterranean. The air became increasingly stable as the ridge over the central Mediterranean developed. Southern Iberia was in stable mT air. The Gibraltar ascent shows a pronounced subsidence inversion between 955 and 925 mbar. Variable amounts of thin cloud were reported in this mT air in which temperatures were some 6° C above those in the central Mediterranean.

The secondary depression over Turkey was the result of cyclogenesis over the Aegean in the preceding twelve hours which retarded the eastward movement of the cold front across the eastern Mediterranean. The secondary depression deepened rapidly to give strong winds and heavy rain. The ascent for Nicosia passed through the frontal mixing

73(b) Upper air diagrams for Mediterranean localities. 1200 GMT, 9 January 1968. Bordeaux (44° 50′ N. 0° 36′ W.), Malta (35° 50′ N. 14° 30′ E.), Gibraltar 36° 7′ N. 5° 22′ W.), Sofia 42° 45′ N. 23° 20′ E.), Nicosia (35° 10′ N. 33° 25′ E.)

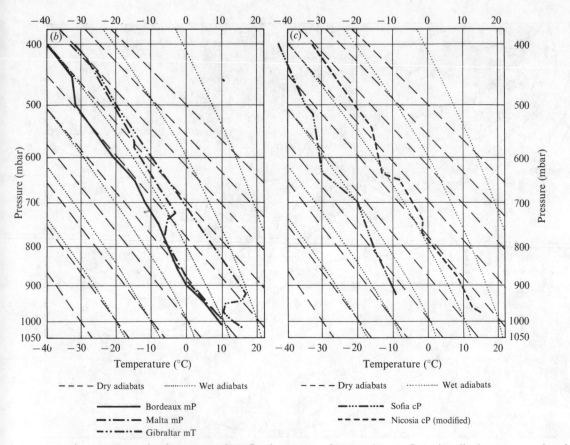

Dry adiabats · · · · · · · · · · Wet adiabats Dry adiabats · · · · · · · · · · Wet adiabats

――――――― Bordeaux mP ―··―··―·· Sofia cP
―·―·―·― Malta mP ― ― ― ― ― Nicosia cP (modified)
···―··― Gibraltar mT

zone into warmer air above 750 mbar. In the rear of these easternmost depressions there was a deep outflow of cP from southern Russia. The Sofia ascent, with a lapse rate less than the S.A.L.R., shows that cP air was convectively stable over the cold continent. Where it passed over the warmer sea, however, intense convection resulted in outbreaks of rain and thunderstorms.

The depression which moved from the Irish Sea to north-east France between 0000 GMT and 1200 GMT was centred over southern Italy at 1200 GMT on 10 January and the cold front of this depression crossed Iberia and the western Mediterranean into north Africa. Behind it mP air penetrated the western Mediterranean. The Bordeaux ascent shows the great depth of this mP air which was unstable to 500 mbar (5400 m) and in which heavy showers occurred. At Gibraltar the mid-day temperature in the mP air on 10 January was 3° c compared with 15° c in the mT air on the previous day.

Continental tropical air from the Sahara is frequently drawn into the circulation of Mediterranean depressions. On the surface it rarely spreads far over the sea. Occasionally, however, outbreaks of cT air take place which penetrate not only to the northern parts of the Mediterranean but also as far north as the British Isles and north Germany to give summer day-time maxima of 30° c and above. This is most likely to occur during April and May ahead of a slow moving cold front off the European coast.

The weather sequence in the winter half of the year normally consists of one to three weeks of cyclonic activity giving disturbed weather, strong winds and periods of rain alternating with about one week of fine, calm weather as anticyclones move east from the Atlantic or south-east from Europe.

The average number of depressions affecting the Mediterranean each year is 76, of which only 7 enter from the Atlantic (Fig. 74). The great majority originate either as frontal wave or Alpine lee depressions over the Gulf of Genoa. The second most important area of cyclogenesis is south of the Atlas mountains. These are mainly lee depressions. The eastern Mediterranean is a minor area of cyclogenesis. The tracks of depressions entering and leaving the Mediterranean are largely

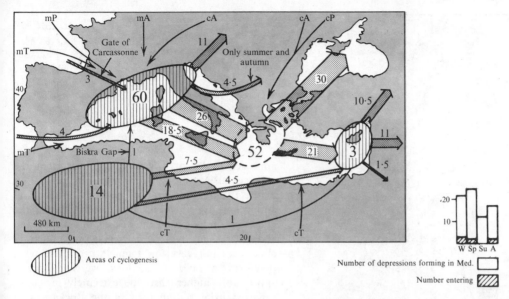

Areas of cyclogenesis

Number of depressions forming in Med.

Number entering

controlled by the major relief features. Some 75% of the depressions leaving the Mediterranean follow tracks into eastern Europe; of the remainder which cross the Middle East some eventually penetrate into the Indo-Gangetic plains of northern India.

Pacific North America

As explained on p. 110, maritime influences extend inland from the Atlantic coast of Europe because of the character of the relief. Only in Scandinavia, where the Kiolen mountains act as a marked barrier to air movements, are maritime conditions confined to a narrow coastal stretch as in western North America where the western Cordilleras lie athwart the northern hemisphere westerlies.

The height of the Pacific coastal ranges, the Cascades (2000–3000 m), and of the inland Sierra Nevada (3000 m +) is such that they act as a barrier to the eastward movement of mP and mT air masses and to the westward movement of cP and cA air masses. They also retard the eastward movement of Pacific frontal systems. As explained on p. 9, the disruption of the upper westerlies by the Cordillera results in the formation of upper ridges of warm air over them, leading to the formation of blocking surface anticyclones. These blocking anticyclones cause many of the Pacific frontal depressions to be steered northwards along the coast into the Gulf of Alaska. To the east of the Rockies upper troughs of cold air frequently occur and lead to cyclogenesis over the interior lowlands.

The temperature conditions experienced at coastal stations show differences from those experienced in the same latitudes on the eastern side of the North Atlantic. This is to some extent due to the difference in temperature of the ocean currents on the eastern side of the North Pacific compared to their counterparts on the eastern side of the North Atlantic. The warm North Pacific Drift flowing polewards along the coast north of 50° N. only extends its influence to the southern coast of Alaska, i.e. to 60° N., due to the N.E.–S.W. orientation of the Alaska peninsula, whereas the North Atlantic Drift penetrates into the Arctic Ocean and influences the temperature conditions as far north as 70° N., including the arctic coastlands of north-west Russia. The North Pacific Drift is also cooler than the North Atlantic Drift. This is especially noticeable in mid-winter when the positive temperature anomaly over the Alaskan Gulf is only 11° C, compared with 28° C off the coast of Norway. The cold Californian current, however, is much colder than the Canaries current and has a marked effect on coastal temperatures south of 40° N., especially in summer when off-shore winds cause upwelling of cold water along the coast.

Winter

The whole of the Pacific coast, with the exception of the Alaskan coast west and north of the panhandle, has a marked concentration of precipitation in the winter half of the year when depressions, steered by the polar frontal jet, are following their most southerly tracks across the Pacific. The polar

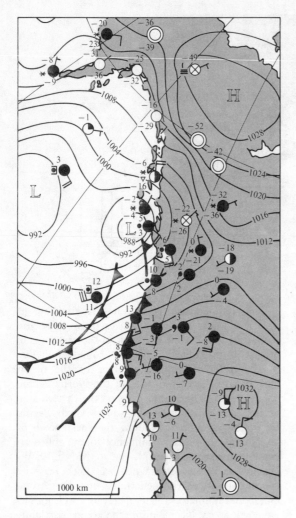

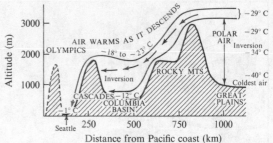

approached the coast very cold cP air from the continental interior was drawn into its circulation and, as the pressure gradient between it and the cold anticyclone over the Mackenzie basin tightened, the cP air came to dominate the weather of both the land and sea areas north of 54° N. to give predominantly clear skies but very low temperatures. The temperatures at coastal stations were, however, appreciably higher than the extremely low ones reported from stations east of the Rockies (as a comparison of temperatures between 55° N. and 60° N. shows). The reason for this is summarised in Fig. 76 which shows that cP air over the Great Plains becomes extremely cold and stable in its lower layers as a result of radiational cooling during the long winter nights and a marked inversion develops in it at a height well below that of the crest of the Rockies. It is the warmer air above the inversion, and especially above the Rocky crest line, which is drawn westwards in situations such as that shown in Fig. 75. As the air descends the western slopes of the mountain ranges it is warmed by compression at the D.A.L.R. to give the higher temperatures experienced in the intermontane basins and along the coast.

Summer

In spring the winter jet stream over California wanes and another develops near the arctic circle. As a result precipitation along much of the Pacific coast gradually declines to a summer minimum as the Pacific depressions follow more northerly tracks. The decline is not interrupted by a secondary spring maximum as in western Europe.

In summer and early autumn Pacific frontal depressions associated with the high latitude jet stream are concentrated north of 60° N. From the Hawaiian anticyclone, whose oceanic axis is about 38° N., a ridge extends north-eastwards along the Pacific coast to north of 50° N. Steady northerly winds on the eastern side of the ridge result in the upwelling of cold water along the Californian coast and cause the effects of the Californian cold current

frontal jet, located north of 60° N. in early autumn, moves steadily south during late autumn and early winter to reach its most southerly position over California in February. Fig. 75 shows an occasion when a depression, which had travelled east-north-east across the Pacific, and its associated cold front, were giving precipitation between 37° N. and 55° N. Unlike those of the east North Atlantic, the warm fronts of the North Pacific are usually weak, and frontal precipitation largely comes from cold and occluded fronts. Their retardation by the coastal ranges causes the precipitation from them to be of long duration. Heavy showers occur in the mP air behind the fronts, especially on the windward slopes of the coastal ranges, the additional uplift given to the unstable air causing the cumulus clouds to grow in depth.

As the depression centred at 50° N. 130° W.

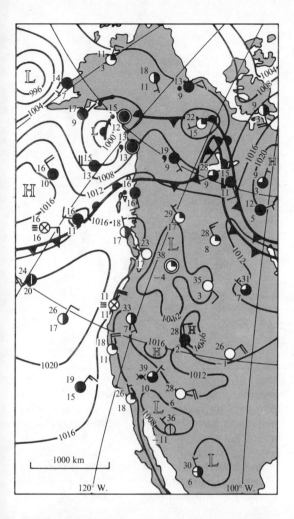

1000 km

120° W.   100° W.

temperatures and dewpoints at inland stations indicate that cT air covered that part of the continent south of the southern-most front. This cT air had been transported poleward from its desert source in south-west U.S.A. and north-west Mexico. Such outbreaks of cT air cause considerable damage to crops by desiccation (cf. the Sukhovey of the U.S.S.R. p. 86).

The mT air circulating round the ridge of high pressure off California was chilled as it crossed the cold Californian current and sea fog, a common occurrence in summer, was reported at 43° N. 125° W. The weak frontal lows, in which the air masses were mP and mA, were giving light precipitation, especially over the coastal mountains north of 55° N.

The southern hemisphere

With the southern Hadley cells (p. 7 being some 5° nearer the equator than their northern counterparts, the southern hemisphere westerlies approach closer to the equator than those of the northern hemisphere. They are also stronger at all heights (see p. 8).

The weather pattern in the greater part of the mid-latitudes in the southern hemisphere is governed by the eastward movement of the subtropical high pressure cells between 20° S. and 40° S. and by the frontal depressions which mainly travel south of 40° S. From the depressions cold fronts trend north-westwards and form troughs between the subtropical high pressure cells which are c. 45° of longitude apart.

In the southern hemisphere the amplitude of the Rossby waves is smaller than in the northern hemisphere and blocking anticyclones rarely develop except over the southern Andes (see p. 8). When anticyclones do form in the westerly air flow the most favoured areas are south of Australia (c. 45° S. 140° E.), south-east of South Africa (c. 40–45° S. 40° E.), between Cape Horn and 62° S. 20° W., the sector 20°–60° S., 100°–120° W., and over the Antarctic continent (c. 75° S. 60° E.). They seldom remain stationary however, but drift eastward.

The absence of large land areas in the middle latitudes of the southern hemisphere means that the zonal thermal pattern is not distorted by cold, continental anticyclones in winter and there are no sources of cP air. Cold air masses only originate over the Antarctic continent and the oceanic waters fringing it. Anticyclonic development is most frequent in eastern Antarctica (60 E.–140 W.) and

on coastal temperatures to be most pronounced at this season. This, combined with the anticyclonic subsidence in the east of Pacific high, results in marked atmospheric stability and is the cause of the low summer rainfall and of the persistent smogs which affect cities such as Los Angeles.

Figure 77 is an example of a summer synoptic chart. The surface high pressure over the Pacific was associated with an upper warm ridge whose axis was c. 150° W. A weak upper trough aligned along the Pacific coast separated this ridge from another pronounced one which dominated the whole of the continent to c. 100° W. with a high pressure centre at 39° N. 116° W. There were only slight variations of sea-level pressure over the western Cordillers south of 55° N. and these were due to the formation of the three weak thermal lows, in which occasional showers occurred, caused by day-time heating. The great differences between air

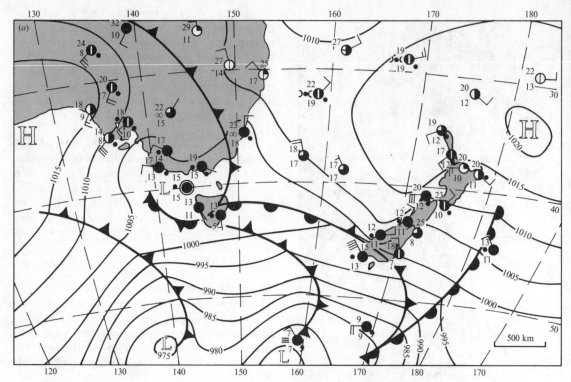

from there cold outbursts sweep into the Indian Ocean sector. Those from the Weddell Sea area (20° W.–60° W.) affect all parts of the South Atlantic and sometimes the Indian Ocean. Such is the depth and frequency of the outbursts from eastern Antarctica that southerly blizzards are common near McMurdo Sound and the western Ross Sea (70–75° S., 160–170° E.) and cold climates are experienced in Kerguelen Island (49° S. 70° E.) and Heard Island (53° S. 73° E.). Mauritius (20° S. 58° E.) on the other hand, has a tropical climate. Therefore in this sector of the Indian Ocean the mid-latitude climates have a small latitudinal extent.

Although outbreaks of antarctic air do penetrate to low latitudes they seldom reach the main land areas, with the exception of the extreme south of South America. The most frequently experienced air masses in southern South America and Australasia are:

Maritime polar, which originates between 55° and 65° S. Winter outbreaks of this cause sudden falls in temperature of 4° to 10° C in southern Australia. Snowfall in the highlands occurs in these outbreaks. In South America the 'pampero' is associated with this air mass.

Maritime subpolar, which originates in the strong westerly wind belt between 40° and 55° S. Its rapid flow over the oceans results in marked uniformity of temperature and humidity. Summer temperatures of 10° to 15° C occur within it.

Maritime subtropical. This originates between 25° and 40° S. Temperatures in it are 15° to 21° C. It has a much higher humidity than maritime subpolar air.

Maritime tropical, which originates equatorwards of 25° S. and gives temperatures of 21° to 27° C.

Continental tropical from source regions in the interior of Australia, South Africa and South America.

Australasia

Summer

In summer the anticyclones in the westerly air flow travelling eastwards between 30° and 45° S. control weather over the most southerly parts of Australia and cause the dry summers experienced there. Their centres are normally *c.* 45° of longitude apart and the cold frontal troughs between them, separating mT air on the western flank of one anticyclone from msP/mP air on the eastern flank of the next one, are

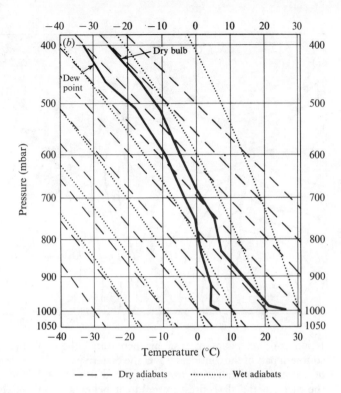

78(*b*) Upper air diagram, Christchurch, New Zealand, (43° 33′ S. 172° 39′ E.), for 0000 GMT, 12 December 1967

— — — Dry adiabats   ⋯⋯⋯ Wet adiabats

usually innocuous. As the weak troughs cross the Murray-Darling basin, however, day-time heating periodically gives rise to the formation of thermal lows which travel south-eastwards along the fronts to give outbreaks of rain over New South Wales. Such outbreaks account for the summer maximum of rainfall in the eastern part of the state. The frontal troughs, linked to depressions whose centres track west–east between 55° S. and 45° S., are more pronounced and active south of 40°/45° S. and give appreciable summer rain in Tasmania and South Island, New Zealand.

Figure 78 shows a summer synoptic situation when the subtropical jet was over South Australia and New Zealand. The North Island, New Zealand, was mainly under the influence of the large, warm anticyclone centred to the north-east of it. Such anticyclones account for the summer minimum of precipitation of North Island. Over Tasmania and neighbouring parts of the mainland a shallow wave depression was giving intermittent rain. Further north over the mainland outbreaks of rain occurred in the msP air behind the cold front and fine weather in the cT air ahead of it. On the western side of South Island, New Zealand, heavy orographic rain was occurring in the strong north westerly air flow, between the Tasmanian depression and the Pacific anticyclone. The deflection of the

air stream by the southern Alps had caused the formation of a high pressure ridge to the west of the island and a low pressure trough to the east. The marked contrast between air temperatures and dewpoints on the western and eastern sides of the island indicate that a pronounced föhn was being experienced in the east. The effect of the drying out of the lower layers by orographic rain in the west and of warming by descent in the east is apparent on the Christchurch upper air ascent (Fig. 78). From 1000 to 990 mbar a super D.A.L.R. of 5° C/ 100 m occurred; between 990 and 830 mbar a D.A.L.R. The approximation of the ascent curve to the S.A.L.R. above 830 mbar shows the general stability of the air mass. The much greater differences between air temperatures and dewpoints below 830 mbar than above was due to the loss of moisture on the windward slopes and föhn warming (see p. 90) on the leeward side.

Winter

In winter the anticyclone cells travel west–east north of 30° S. and mid-latitude depressions, which are much more vigorous in this season, also follow lower latitudinal tracks than in summer. Fig. 79 shows such a synoptic situation. In the troughs behind the anticyclonic cells crossing central Australia cold fronts gave narrow belts of precipi-

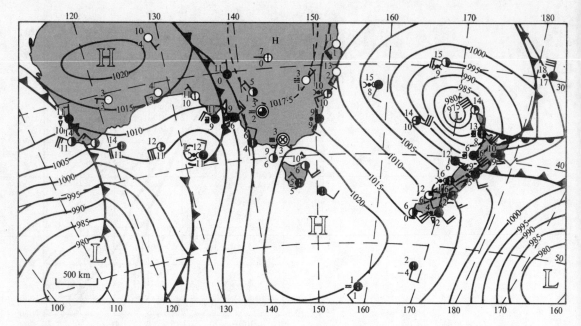

tation and incursions of msP air penetrated the southern part of the continent. Over the eastern part of the mainland radiation fog had formed during the night in the slack pressure gradient below an upper ridge, and was persisting in several places at the time the observations were made (1000 hours sun-time at 150° E.). The anticyclone south of Tasmania had formed below an upper cut-off high. To the east of the upper ridge a large cut-off low was located over the Tasman Sea and the deep surface depression associated with it had a diameter of *c*. 3000 km. Showers and outbreaks of rain were occurring in the unstable polar air over the Tasman Sea, the New South Wales coast and on the eastern side of New Zealand. Föhn conditions were being experienced on the western side of South Island, New Zealand.

The winter maximum of precipitation in the southern margins of Australia and North Island, New Zealand is due to the more frequent passage of frontal depressions at that time of the year. Seasonal contrasts in precipitation are less marked in South Island, which is in the track of frontal depressions throughout the year.

The weather sequence in frontal troughs crossing New Zealand, especially South Island, varies according to the trend of the isobars and wind direction behind the front. North-westerlies ahead of the fronts give thick masses of cloud and orographic rain on the western slopes of the mountains, and comparatively clear skies to the east. With west-south-west to north-west winds behind the cold front the west coast receives heavy frontal rain followed by showers; east of the mountains the passage of the front is barely noticeable. With south-south-west to south winds behind the front frontal rain occurs on both coasts. South-easterly winds give a northward moving line of Cb cloud on the east coast and clear skies on the west.

9 Mid-latitudes – continental interior and east coastal areas

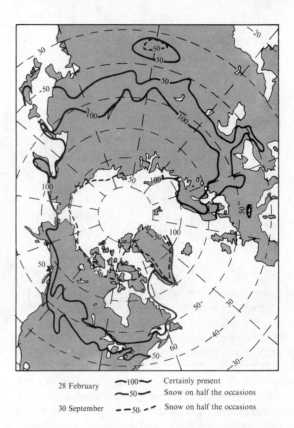

28 February   —100—   Certainly present
            —50—   Snow on half the occasions
30 September   - -50- -   Snow on half the occasions

In chapter 5 it was explained that the permanently ice- and snow-covered areas of the antarctic and arctic regions are source regions of arctic air throughout the year. In the northern hemisphere the sources of very cold air masses extend southwards in winter to include large parts of North America and Eurasia. This follows the rapid fall of ground temperatures in autumn through radiational cooling in those areas unaffected by maritime influences. This is very quickly revealed in the air temperatures. Thus at Winnipeg mean daily minimum temperatures fall below 0° c in mid-October and remain so until the end of April, whilst mean daily maximum temperatures are below 0° c from mid-November to the end of March. The probability of fallen snow lying on the ground increases greatly once the mean daily minimum temperature is below 0° c and once the mean daily maximum is below this level the chances of snow melt are slight. Fig. 80 shows the probability of snow lying on 30 September and 28 February. At the end of September there is 100% probability of snow only on Ellesmere Island to the north-west of Canada, but a 50% probability (i.e. new snow lying on half the occasions) over Greenland, most of the islands north of Canada and the greater part of the sector of the Arctic Ocean between 80° E. and 20° W. The probability is lower than 50% in the sector of the arctic between 20° W. and 80° E. because of the effect of the warm waters of the North Atlantic Drift.

As winter progresses the probability of snow falling and lying over large parts of the continental land masses increases. By the end of February there is 100% probability of snow over most of Scandinavia north of 60° N. and over most of the Soviet Union north of 45°–50° N. – an area of some 10 million square kilometres. In North America, most of Canada and Alaska, an area of over 5 million square kilometres also has a 100% probability of snow lying. The areas with a 50% probability of snow lying are even more extensive in Eurasia and include parts of eastern Turkey, much of China

north of c. 42°–45° N. and the Tibet plateau. In central and eastern North America the 50% probability extends south to similar latitudes. By mid-winter then an area of over 25 million square kilometres of the northern Hemisphere is likely to be snow covered, and the northern continental interiors of Eurasia and North America are source regions of cP air masses. The mid-latitude westerly air flow is responsible for the transportation of the cP air to the eastern margins of North America and Asia, where it dominates the weather throughout the winter north of c. 40° N.

Eurasia

Winter

The lowest winter temperatures experienced in mid-latitudes occur in the interior of Asia north of the Himalayas. Fig. 81(a) is an example of a winter synoptic weather map for the Soviet Union which shows the very low temperatures and high (sea level) pressures which result from the intense radiational cooling of the lower layers of the troposphere. It must be borne in mind in studying

81 (*a*) Synoptic chart, 0000 GMT, 24 December 1966

81 (*b*) (*below*) Upper air diagrams for Novosibirsk (55° 0′ N. 83° 5′ E.) (1) and Yakutsk (62° 5′ N. 129° 40′ E.) (2), 0000 GMT, 24 December 1966

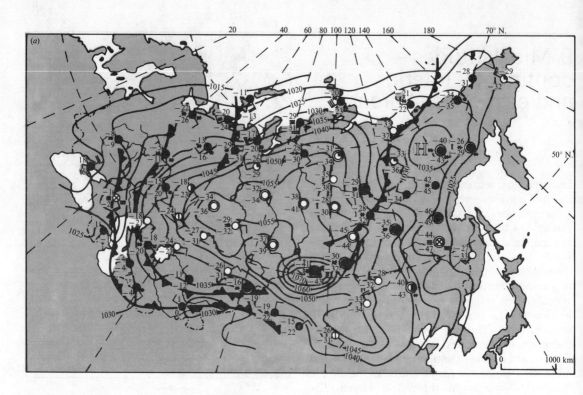

this map that the reports span a period of sun-time ranging from 0200 hours at 30° E. to 1200 hours at 180° and only the area approximately south-east of a line from 66½° N. 180° to 46° N. 120° E. would be experiencing daylight. The 1015 mbar isobar is normally taken as the boundary between high and low pressure and in this instance virtually the whole of the Soviet Union and parts of adjoining areas had a sea-level pressure exceeding this. The general anticyclonic curvature of the isobars emphasised the dominance of the Siberian anticyclone. The main anticyclone centre at 50° N. 95° E. is the 'normal' position, but the central pressure was unusually high and close to the upper limit of recorded sea level pressure. A weak frontal low separated the main anticyclone from a minor one centred over north and east Siberia. In the south-west of the U.S.S.R. warmer air from the Mediterranean was giving warm weather along the Black Sea shorelands. The quasi-stationary front extending from 46° N. 85° E. to south of the Aral Sea through the north-east of the Caspian Sea to 60° N. 38° E. marked the surface boundary between cP air circulating round the south and west of the Siberian anticyclone and air masses of Atlantic origin which had been modified in their passage over the cold land surface. Light snow was falling in this frontal zone in a number of localities. In the northern parts

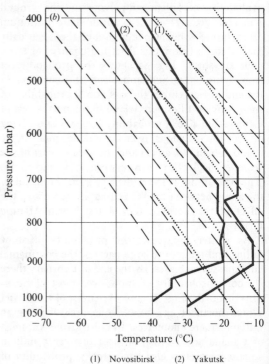

(1) Novosibirsk   (2) Yakutsk

of the U.S.S.R. the non-uniform character of the air masses was shown by the two frontal troughs, each with outbreaks of slight continuous snow along them. The one in eastern Siberia was the most marked with temperatures some 5° to 7° C higher in the less cold air to the west than in the very cold air to the east. The snowfall reported away from the fronts came from thin layer clouds and was light in every case. The lowest temperatures reported were from stations located in valley bottoms in the mountainous and hilly country in the east and south-east of Siberia.

The upper air ascents for selected stations (Fig. 81(*b*)) in Siberia illustrates the coldness of the air at all levels, the highest recorded temperature being $-12°$ C between 900 and 850 mbar above Novosibirsk. The inversion tops at *c*. 900 mbar marked the upper limit of radiational cooling and the base of the warmer, subsiding air above. This occupied the almost isothermal layer between *c*. 900 mbar and *c*. 700 mbar. Above 700 mbar the lapse rate varied between 6° C/km and 4.5° C/km, emphasising the stability of the air at all levels.

The upper winds east of the Siberian front were northerly revealing a flow of arctic air. The temperatures at all levels in this arctic air (Yakutsk ascent) were some 10° C lower than those in the cP air (Novosibirsk ascent).

The Siberian anticyclone is not always as intense as in Fig. 81(*a*). When it covers a more restricted area to the east of the Urals, Atlantic and Mediterranean depressions can penetrate the west and north-west of the Soviet Union. The maritime air masses in these depressions give milder spells of winter weather and periods of continuous snow which sometimes occur as intense blizzards. Depressions which penetrate from the Atlantic have usually crossed the north European plain and the southern Baltic and then tracked north-eastwards across northern European Russia. In their rear outbreaks of arctic air occur and sometimes penetrate to the Black Sea and beyond. Similar depressions give spells of disturbed weather on the Pacific seaboard of the U.S.S.R. Over most of the continental interior, however, the Siberian anticyclone dominates the weather from October to March–April, a period which is mainly calm but intensely cold. In some winters the development of intense blocking anticyclones over Scandinavia results in western Europe being under the influence of cP air for periods of varying length (pp. oo–o).

On the eastern side of Eurasia this dominance is a yearly occurrence, the outward flow of cP air at the surface occurring between the Siberian anticyclone, which begins to develop in October, and the semi-permanent Aleutian low pressure system. This is the winter monsoon which by January has extended its influence over the whole of mainland east Asia. The outbursts of cP air from Mongolia, each preceded by a cold front, occur at intervals of about a week. These fronts only extend 1500–2000 m above ground level and give little precipitation over the continent since the air is very cold, dry and stable. Their southward progress across China is frequently retarded by such west–east hill ranges as the Chinling Shan and Nanling Shan. Cyclogenesis however often takes place on such fronts in the Yangtze valley region, which is the zone of convergence between the two branches of the upper westerly air stream flowing round the northern and southern flanks of the Tibet plateau. North-westerly winds are characteristic of the upper air flow over north China, south-east Siberia and north Japan and south-westerly ones over south China and east China Sea. The subtropical jet stream in this south-westerly air flow is markedly constant in position, being orientated from *c*. 22° N. 100° E. to 32° N. 140° E., and by rapidly removing air aloft it stimulates the upward ascent of air below it. This in turn stimulates cyclogenesis on trailing cold fronts in the Yangtze valley area and the rejuvenation of mid-latitude depressions which have travelled in the air flow across the Indo-gangetic plain. The zone of maximum winter precipitation in China lies along this jet axis.

Figure 82(*a*) shows the synoptic situation over east Asia on 9 January 1968. Over Mongolia and neighbouring parts of Siberia and China the main Siberian anticyclone gave fine or fair weather. From Mongolia and south-east Siberia a strong outburst of cP air had taken place as the deep depression centred to the south-east of the Kamchatka peninsula travelled on a cyclonically curved track across southern Japan from the mouth of the Yangtze where it formed on 5 January. On the same day a secondary depression formed near south-west Japan and tracked north-eastwards across the country. The two depressions merged on 9 January to produce the very deep, intense depression shown. During the four days central pressure fell from 1016 to 960 mbar.

The Chita ascent (Fig. 82(*b*)) shows that the air was stable, due to the difference of *c*. 5° C between the air temperature and the dewpoint although the lapse rate approximated to the S.A.L.R., up to 725 mbar. Above this level the difference between

82(*a*) Synoptic chart, 1200 GMT, 9 January 1968

82(*b*) (*below*) Upper air diagram, Nanning (22° 50′N. 107° 8′E.) and Chita (52° 15′N. 113° 15′E.)

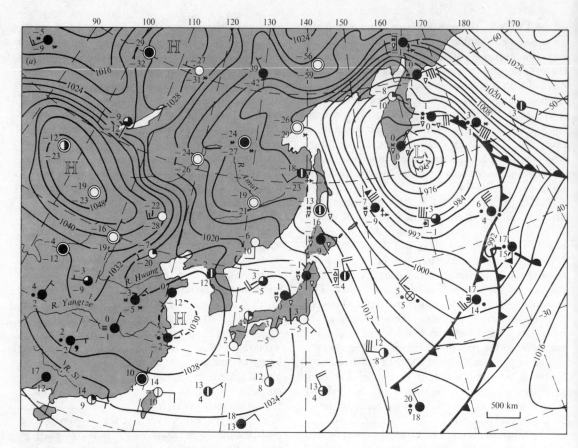

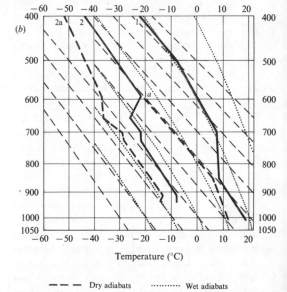

the air temperature and dewpoint was even greater and there was a marked subsidence inversion between 647 mbar and 587 mbar. Hence the clear skies over south-east Siberia and Manchuria. Once the cP air left the continent the temperatures and moisture content of the lower layers increased as it crossed the Sea of Japan. It was now convectively unstable and the additional uplift provided by the western slopes of the mountains caused snow showers to occur on the west side of northern Japan. Such shower activity is the cause of the winter maximum of precipitation on the western side of north Honshu. A short period of calmer, warmer weather occurs on the Pacific side of Japan between each outbreak and is associated with the passage of a ridge of high pressure. In this instance a ridge of high pressure from the Siberian anticyclone had extended its influence over south-west Japan, the east China Sea and the whole of China. Over China temperatures below 0° C were being experienced as far south as 30° N. (sun-time 2000 hours at 180° E.). South of *c.* 30° N. the cP air, warmed and moistened in its long passage across the east China Sea, gave milder, cloudy weather with outbreaks of light rain

Temperature (°C)

– – – Dry adiabats ·········· Wet adiabats

1 Nanning ascent
1a Nanning dew points
2 Chita ascent
2a Chita dew point

and drizzle from low stratus and stratocumulus clouds.

At higher levels in the troposphere the upper westerlies were dominant over the whole of east Asia. The marked temperature difference between the upper air which had reached southern China from south of the Tibet plateau and that which had travelled round the north of it can be appreciated by comparing the upper air ascent of Nanning with that for Chita (Fig. 82(b)). On the Nanning ascent the almost isothermal layer between 850 and 700 mbar shows the transition between the low level modified cP air of Siberian origin and the higher level air from northern India. The low humidity of this air above 850 mbar is shown by the great difference between dry bulb and dewpoint temperatures. The marked stability of the air is also apparent. (Note Figs. 73(a) and 82(a) are synoptic charts for the same time and day.)

Summer

In March and April the Siberian anticyclone gradually weakens and outbreaks of cP air occur less frequently. In the west Atlantic and Mediterranean air masses and frontal depressions penetrate further into the continent. In the east incursions of warm, moist, tropical air, in which anticyclones develop, occur over China. These move eastwards between 20° N. and 45° N., giving spells of calm, settled weather over much of China and Japan. In coastal areas the cooling of outbreaks of warm air by the still relatively cold waters results in widespread occurrences of sea fog and low stratus.

The heating of the continent in summer causes pressure to become generally low, but with an overall synoptic pattern of weak anticyclones and shallow depressions over the continental interior and a semi-permanent thermal low over south-east China. In early summer the subtropical high over the western Pacific moves north until the mean position of its centre is *c*. 40° N. 150° W. and a semi-permanent cold anticyclone develops over the cool waters of the Sea of Okhotsk. Fig. 83 illustrates these general pressure conditions and the weather associated with them, such as the thunder showers which develop in the thermal lows (e.g. at 55° N. 97° E.) and the low temperatures in the sea fog which commonly occurs over the Sea of Okhotsk.

At the beginning of summer the southerly jet stream moves suddenly to a position north of the Tibetan plateau from its winter position to the south of it. The northward movement of the southern jet is followed by the rapid spread of a south-westerly current of mT air (equatorial), from the Bay of Bengal across south-east Asia, south China and Japan. This air stream can be identified over the South China Seas in Fig. 83. At higher levels the equatorial easterlies normally occur over south China, but were further south of this occasion. Cool mP air from the Okhotsk anticyclone had spread southwards and interacted with mT from the Bay of Bengal and from the Pacific anticyclone to produce the frontal systems extending north-eastwards from south China across Japan shown in Fig. 83. This frontal system gives the *Maiu* rains over China and the *Baiu* rains over Japan. As the chart shows, small waves, some of which have small depressions associated with them, develop on the front and track north-eastwards and are mainly responsible for the summer maximum precipitation in south and central China and much of Japan. Mid-summer variability of rainfall is linked to the strength and position of the *Baiu* front. If it is weak and far north, north China and Japan receive meagre rainfall; if well south a cool, rainy summer results.

The typhoons which had formed on the southern side of the Pacific anticyclone, and deepened as they travelled westwards, were now moving north-eastwards round the western end of the high.

The average number of tropical cyclones within the area 5° N. to 30° N. and 105° E. to 150° E. in the period 1884–1953 was:

J.	F.	M.	A.	M.	J.	
0.3	0.1	0.1	0.3	1.0	1.5	
Jy.	A.	S.	O.	N.	D.	Year
3.8	4.4	4.4	3.0	2.1	0.9	21.9

The damage done in coastal areas by typhoons is partly due to the very strong winds and partly due to the storm surge of the sea when high tide levels are as much as 2 m above normal. Thus at Hong Kong in 1962 typhoon Wanda resulted in 36 ocean-going vessels being sunk or driven ashore, 1297 junks sunk and another 756 damaged, whilst 183 people were killed or missing.

Typhoons also contribute appreciably to the summer rainfall of those areas in their tracks. Thus at Kagoshima in south-west Japan typhoons contribute about 11% of the annual rainfall.

During July the southerly jet disappears, the Okhotsk anticyclone weakens and the Pacific subtropical high moves to its most northerly position. The *Baiu* season then comes to an end. Weak frontal lows, however, still give precipitation in the latitudes of Japan. By late summer the south-easterly monsoon has become established over the China Seas, China and south Japan and a ridge of high pressure frequently extends over the area from the Pacific anticyclone. The increased stability of the air accounts for August generally having a lower rainfall than July. Late summer/early autumn is the period of maximum occurrence of typhoons.

Between late August and early October the polar jet and its associated front moves southwards and the subtropical high weakens. As the front moves southwards the *Shurin* rainy period is experienced over Japan and neighbouring areas to give a second maximum of precipitation. The front marks the boundary between mP air from the re-established Okhotsk high and mT air from the Pacific high. To the north-west of the front anticyclones develop and give periods of settled weather similar to that experienced in late spring.

At this period temperatures are beginning to fall rapidly over Siberia and the Siberian cold anticyclone begins to re-establish itself. As this develops the pressure gradient between it and the Pacific frontal depressions tightens and surges of cold cP air result in the onset of winter monsoon.

North America

On p. 59 it was explained that the Sierra Nevada, Cascades and coastal ranges of British Columbia form the approximate boundary between the west coast maritime and Mediterranean climates and the continental climates of North America.

The air masses which influence the weather east of the Rockies are (Figs. 34 and 84):

Arctic air from the source region which includes Greenland and the Canadian archipelago as well as the Arctic Ocean.

Maritime polar from two sources (*a*) the Gulf of Alaska (Pacific); (*b*) between Greenland and Labrador (Atlantic).

Maritime tropical from the Bermudan anticyclone (Atlantic).

Continental polar which originates over the interior of northern Canada (and by some is considered to be modified arctic air) and the intermontane basins of the west Cordillera in winter.

Continental tropical, which has its source in the arid intermontane basins of the west Cordillera in summer and forms from mT air (Pacific) subsiding on the eastern side of the Hawaiian anticyclone.

The north to south trend of the relief features facilitates the southward spread of arctic and continental polar air masses from north Canada and the northward movement of mT air from the Atlantic and Gulf of Mexico and cT from the intermontane basins.

Pacific air masses are less frequently experienced east of the Rockies, partly due to the barriers to west to east movement in the lower layers of the atmosphere provided by the west Cordilleran mountain ranges and partly due to the frequent development of high level warm ridges and surface

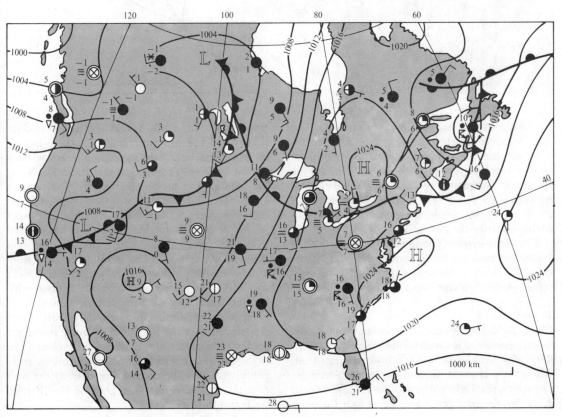

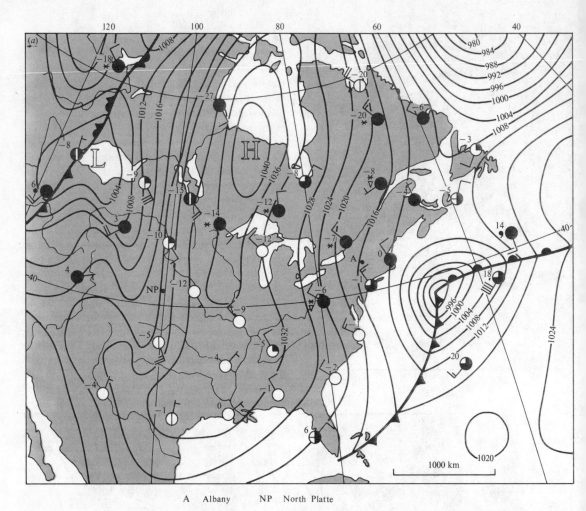

A   Albany   NP   North Platte

blocking anticyclones over the west Cordillera. The most frequent penetrations of Pacific air occur in autumn when there is a predominantly latitudinal air flow over the greater part of the continent. At the surface it commonly reaches the Atlantic coast between *c.* 40° N. and 57° N., being separated from arctic air by the arctic front and from the mT (Atlantic) air by the Pacific front.

The temperature/humidity characteristics of Pacific air masses are very much altered, particularly as a result of the loss of moisture by precipitation, in their passage over the west Cordillera.

Winter

Figure 85 shows an autumn situation when Pacific air masses crossed the west Cordillera on the western flank of a weak upper ridge whose axis ran from Mexico to Hudson Bay. A west-south-westerly upper air stream affected most of the continent north of 30° N., the strongest winds being near the polar

front. East of *c.* 90° W. a weak surface anticyclone occurred below an upper ridge. In the relatively clear sky and light wind conditions near the main anticyclonic centre, night-time radiational cooling had caused fog to develop in the St Lawrence valley. To the west a shallow frontal depression was crossing central Canada. On the trailing cold front of this depression another frontal low had formed over the Great Basin. Little precipitation was reported along the fronts. Behind the cold front a ridge of high pressure limited precipitation in the cool mP air to the occasional coastal or orographic shower. Thundery showers were occurring, however, over the Mississippi lowlands and the south Appalachians in the mT air circulating round the south-western part of the Azores anticyclone. The high temperatures and dewpoints of this air mass contrasted with the much lower ones to the west over the west Cordillera and north of the (Pacific) polar front.

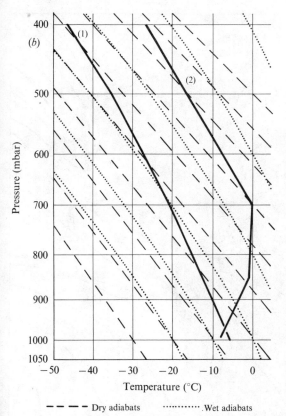

(b)

(1)

(2)

Pressure (mbar)

Temperature (°C)

- - - - Dry adiabats ....Wet adiabats

(1) Albany 42° 49′ N. 73° 49′ W.

(2) North Platte 41° 09′ N. 100° 45′ W.

As autumn progresses the temperatures in north Canada fall rapidly with the decreasing amount and intensity of insolation. The arctic front moves southwards to a mean position along the southern boundary of the Boreal forest and cold anticyclones increasingly develop in the arctic air mass over the interior of Canada. Periodically they intensify to dominate the weather over the greater part of the continent.

This happened in mid-December 1968 (Fig. 86(a)). Most of eastern and central Canada was covered by arctic air with the arctic front, running from Labrador to just south of Winnipeg, separating it from the cP air to the south. The temperature contrast between the two air masses was marked. The meridionally elongated anticyclone shown moved slowly south-eastwards and was a source region of cP air in which temperatures were at or below freezing point throughout eastern North America, except in the Florida peninsula.

The upper air ascents in Fig. 86(b) show the west-east changes in the upper air temperatures in the anticyclone near 40° N. The contrast between warmer air on the western side (North Platte) and colder air on the eastern side (Albany) is marked. At North Platte the cP air only extended to the 850 mbar level, milder Pacific air occurring above.

The deep depression off the Atlantic coast in Fig. 86 formed 24 hours previously over Georgia at the tip of a polar frontal wave. In the 24 hours the centre travelled c. 1700 km and pressure fell 18 mbar. Twenty-four hours later it was centred over south-west Newfoundland with a central pressure of 968 mbar. Such intense depressions along the polar front between the cP and mT (Atlantic) air masses account for the appreciable winter precipitation in the Atlantic states. Fig. 87 shows the situation four days later. The anticyclone had moved south-east to an offshore position and from it an elongated ridge still dominated the weather of the Atlantic and eastern Gulf states, where temperatures were still below freezing point. Over the Atlantic the warming of the lower layers of the arctic air by the warm waters is evidenced by the reported temperatures. mT air on the southern flank of the anticyclone had penetrated the Mississippi lowlands and cyclogenesis has resulted in the deep depression centred at 40° N. 100° W. which now dominated the weather of the continental interior south of 51° N. The temperatures in the warm sector were up to 20° C higher than those recorded four days earlier in the same localities. Such marked temperature fluctuations are a feature of the winter weather south of c. 45° N. and east of 100° W., beyond which limits mT air rarely penetrates in winter. The cP air sweeping southwards in the rear of the depression heralded another cold spell.

Blizzards are the greatest winter weather hazard east of the Rockies. They are snowstorms lasting at least 6 hours, during which temperatures fall below − 12° C, winds exceed 39 km/h and visibility is less than 800 m. Synoptic situations such as that shown in Fig. 87 are the main cause of blizzards. One of the worst experienced over the northern Great Plains of the U.S.A. occurred between 2 and 5 March 1966 and in some places it lasted the whole four days. This blizzard followed the formation of a small depression over south-west Montana on 1 March. Its subsequent track and central pressure are shown on Fig. 88. As it moved south-eastwards on 1 March cold arctic air spread southwards in its rear and snowfall became general over most of Montana, Wyoming, Nevada, South Dakota and southern North Dakota. On 2–3 March an upper trough from the Pacific crossed the western states

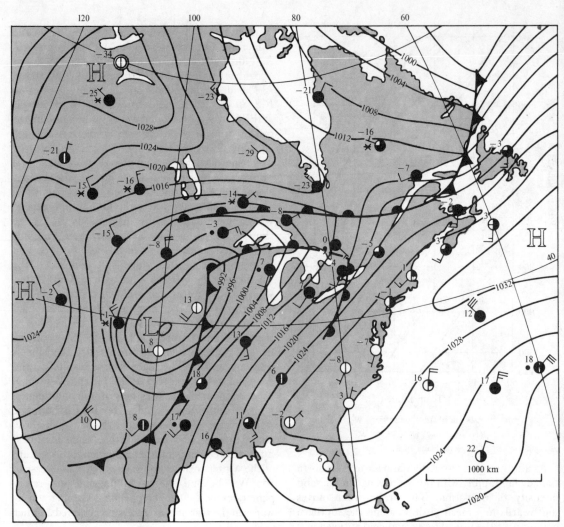

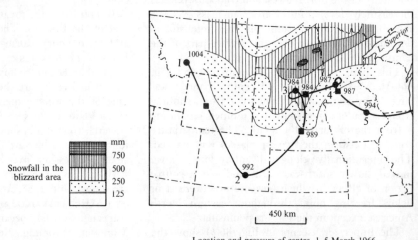

Snowfall in the blizzard area

mm
750
500
250
125

Location and pressure of centre, 1–5 March 1966

● 0600 GMT   ■ 1800 GMT

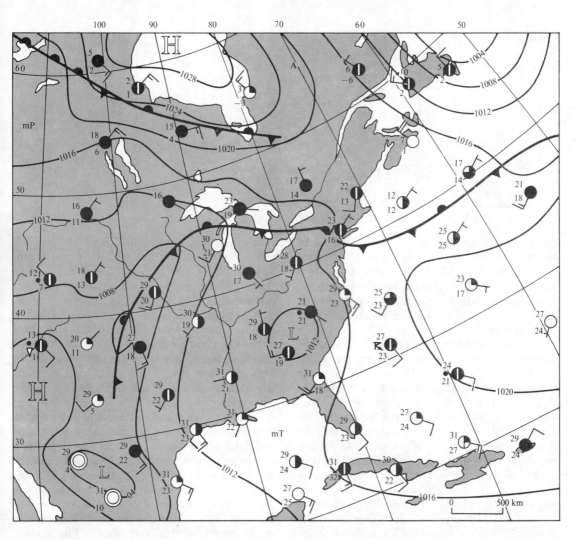

and an upper cut-off low formed in it, and deepened. As it moved slowly over the Great Plains the surface depression became associated with it, deepened still further and also moved slowly on 3–4 March, when the affected area was 1000 × 400 km. Winds near the centre gusted to 120–40 km/h during this period. Fig. 88 shows the depth of snow recorded at official stations during the blizzards. In the strong winds, however, considerable drifting took place, the drifts reaching heights of 10–13 m in North Dakota.

During the four days of the blizzard everyday life was greatly disrupted in the northern Great Plains. All transport was halted by the second day of the storm, including three trans-continental trains which were trapped in cuttings. Power supplies were cut, aircraft hangars collapsed, schools closed and newspapers ceased publication. In the Dakotas and Nebraska 74,500 cattle and 54,000

sheep died, as well as other domestic animals. Fortunately, due to the timely weather forecasts, only 18 people died as a direct result of the blizzard.

Summer

During April and May the wedge of Pacific air over the Great Plains is normally pinched out between the arctic and mT (Atlantic) air streams which are juxtaposed east of 100° W. and between 40° and 45° N. Since these air streams are the ones with the greatest temperature and humidity contrasts the Middle West has violent weather in April and May. Since there is no arctic-continental polar confluence in the central Canadian subarctic the weather there is fine but still very cold.

As summer progresses the arctic front retreats to its mean northerly position along the tundra-boreal forest boundary. A wedge of Pacific mP air re-

75

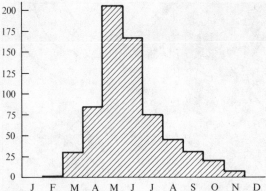

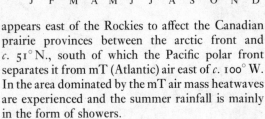

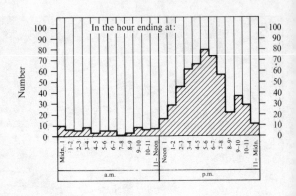

appears east of the Rockies to affect the Canadian prairie provinces between the arctic front and *c.* 51°N., south of which the Pacific polar front separates it from mT (Atlantic) air east of *c.* 100° W. In the area dominated by the mT air mass heatwaves are experienced and the summer rainfall is mainly in the form of showers.

Figure 89 illustrates a summer heatwave situation in an mT air mass covering the greater part of the U.S.A. east of the Rockies. The western end of the Atlantic subtropical high extended over the eastern U.S.A. and hot, moist air from the Carribean and the Gulf of Mexico was circulating round the southern and western sides of it. At the time of the observations (1200 hours sun-time 90° W.) maximum day-time temperatures had not yet been reached over the eastern states, but temperatures as high as 30° c were being recorded as far north as the southern Great Lakes. A thermal low had formed over the southern Appalachians. That instability occurred within the area of the low is shown by the station at *c.* 43°N. 80°W. reporting rain.

As frequently happens in summer, another thermal low was present over the Mexican–U.S.A. borderlands, but this did not extend to any great height. mP (Pacific) air covered the prairie provinces and the Great Plains. The low temperatures being recorded in the vicinity of Hudson Bay and the Labrador–Newfoundland–maritime provinces region occurred in arctic air in which the cold anticyclone centred over Hudson Bay had formed.

Tornadoes are a common occurrence in the United States east of the Rockies in summer, affecting especially the Middle West and the Gulf states. They are associated with vigorous Cb clouds in which the continually evolving individual cells which normally feed a Cb cloud are superceded at the mature stage by a much larger super-cell in which intense up-draughts and down-draughts co-

exist in a nearly steady state. This happens when there is a strong vertical windshear and warm, humid, low level air is trapped below cold dry air aloft. The air in the warm up-draught rotates cyclonically and its buoyancy is maintained by the release of latent heat of condensation. From the top of the up-draught precipitation particles are carried by the divergent air flow towards the periphery of the cloud, but mainly in the direction to which the high level winds are blowing. If the cold air inflow at middle levels is dry, it is further cooled by the absorption of latent heat of evaporation where it enters an area of falling precipitation. The negative buoyancy so caused results in the cold down-draught at the rear of the storm. The tornadoes originate in the cyclonic shear where the cold air undercuts the foot of the warm up-draught. At the time of formation the funnel-shaped cloud is more or less vertical. The upper part of the whirl becomes slanted and sometimes detached as the Cb cloud is moving at the speed of mid-level winds. The average diameter of tornadoes is 250 m, but the destructive path may exceed a width of 1 km. With the passage of a tornado pressure falls suddenly by about 25 mbar and buildings, inside which pressure does not fall, explode as it passes over. Wind speeds in the cyclonic circulation of the core have been estimated at around 350 km/h.

In the period 1916–61, about 11,053 tornadoes were reported in the United States. One of the worst affected states was Iowa where 667 were recorded, of which 68 had paths of 40 km or more. The two longest paths were of 288 km. All but 4% of the tornadoes moved from between south and north-west; the most common direction of travel, followed by 56% of them, was from south-west to north-east. As Fig. 90 shows, May was the month with the greatest number and, as stated on p. 75, this, along with April, is in the period when violent weather can be expected in the Middle West, due to the direct interaction of arctic and tropical air masses. The majority of tornadoes form on squall lines or severe cold fronts.

Figure 91 shows the time of occurrence of tornadoes in Iowa. They occur most frequently in mid or late afternoon when convectional overturning of the lower layer of the atmosphere is at a maximum.

10 High latitude areas

Of great importance in considering the high latitude climates is the contrast between south and north. Antarctica is a high compact continent with an average elevation of 2000 m. It is 10 million sq km in area and is surrounded by an unbroken ring of ocean, narrowest at Drake Passage between Grahamland and South America, where it is constricted to about 1000 km. Elsewhere the ocean is from 2400 km to 3200 km in width. With its maximum extent of sea-ice attached, Antarctica virtually becomes a single snow and ice surface of 25 million sq km. Around the north pole is a deep ocean of about the same size as the antarctic continent. It is surrounded by land but not completely so. Warm Atlantic water can obtain access through the considerable gap between Greenland and Scandinavia. Passages west of Greenland through the Canadian islands are more difficult to negotiate and the Bering entrance north of the Pacific is narrow. In summer the Arctic pack-ice normally withdraws from the shores of the U.S.S.R. and most of mainland Canada, though not from Greenland and all the Canadian islands. It contracts to about 8 million sq km. In winter it joins the continents to give a vast ice- and snow-covered area of 40 million sq km, which is equal to about 10–11% of the global surface. At this season it is a potent source of cold air masses.

Temperature
Winter conditions at either end of the globe tend to be more similar than summer conditions though at both seasons the greater elevation at the south pole is an important factor in giving lower temperatures there.

Figure 92 shows an intense anticyclone over the Arctic Ocean in February 1968. Every station shows a sub-zero temperature, except for Reykjavik (Iceland) with $+1°$ C. Temperatures in the area between Iceland and Novaya Zemlya, which was under the influence of the depression south of Spitzbergen, while low, are 20° higher than the air mass surrounding the large high north of the Bering Strait. The lowest temperature recorded is over Siberia, though northern Greenland, northern Canada and a station on the arctic pack-ice record temperatures of nearly 40° c below zero.

The July situation (Fig. 93) shows a tremendous change in temperature. Drizzle is reported from a drifting ice-island camp at 86° N. 160° E. and there is only one sub-zero temperature on the chart. Differences in temperature around the Arctic Ocean are much smaller on this chart than in winter. Inland in Siberia, but still north of the arctic circle, particularly high temperatures can be noted at Verkhoyansk, 19° c, and Olenek, 29° c, both seven hours after local noon. The effect of the long hours of daylight must be remembered. Rain and drizzle are not uncommon at the north pole in summer. At the south pole in the summer of 1967–8, −18° c was the highest temperature recorded. For Vostok the recorded *maximum* is −21° c. Along the east antarctic coast which reaches the polar circle +8° c is the highest recorded.

Everywhere in high latitudes long summer daylight and long winter nights are important factors. Poleward of the circles all places have one midnight sun and one 24-hour night.

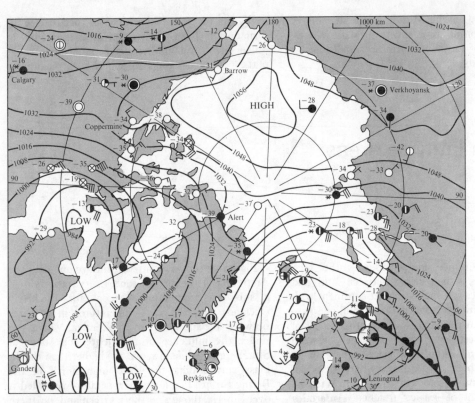

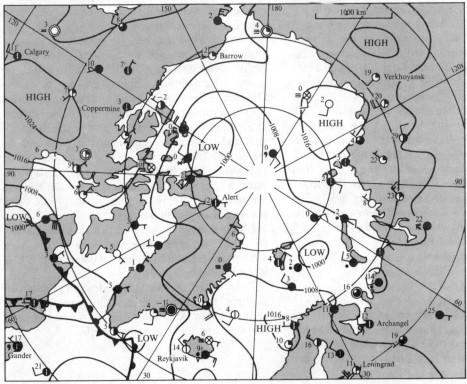

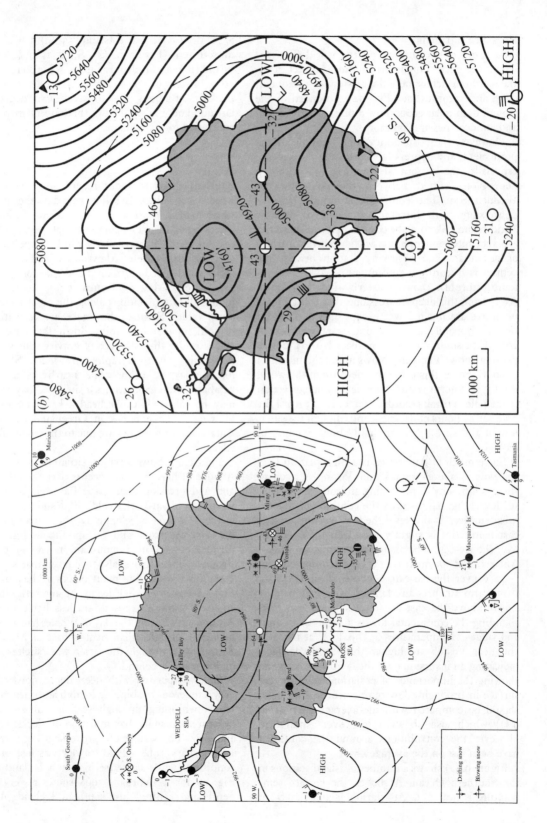

94(*a*) Synoptic chart. 1200 GMT. 25 April 1958

94(*b*) 500 mbar chart. 1200 GMT. 25 April 1958

Pressure patterns

The graph of pressure round the world (Fig. 20) shows the deep trough surrounding Antarctica. In the southern hemisphere depressions are frequent, deep and vigorous (see Fig. 94). Poleward of the trough both polar areas show a pressure maximum and anticyclones are common.

However, the polar anticyclone is by no means permanent, and depressions do invade the ice-cap, especially the eastern sector. Fig. 94 shows a depression centred between the pole and the Weddell Sea. The continent is ringed by four depressions with intervening ridges or shallow anticyclones. Skirting the coast of eastern Antarctica is a very intense depression with a central pressure of 944 mbar. The observing station nearest the centre is recording continuous heavy snow, a temperature of 0° c and an easterly wind of 150 km/h. Away from this deep depression other coastal temperatures are much lower except on the northern tip of Grahamland where 0° c is also recorded. Inland the south polar station records $-64°$ c with drifting snow. The wind blows along the meridian of 90° E. but as this is the southern tip of the world it must be a north wind – there is only one direction at the pole. Vostok records $-69°$ c. By 25 April the pole has 24 hours of darkness and, though the sun is low in the sky, 80° S. has still about 8 hours of daylight.

Since much of Antarctica is a high plateau (e.g. the pole is about 3000 m and Vostok is 3500 m) there is difficulty in producing an isobaric chart for sea level. Fig. 94(b) shows the contours of the 500 mbar surface. At this level the low over the coast of east Antarctica is slightly less deep than the low over the ice-cap. The air in the depression off east Antarctica at 500 mbar is $-32°$ c, as compared with $-42°$ c over the ice-cap. Since cold air is denser than warm air, pressure falls off with height more slowly above the east antarctic surface low. Surrounding the continental low is a series of troughs and ridges. Temperatures at this level at the pole and near Vostok are higher than at the surface, indicating an inversion over the central ice-cap and showing the importance of radiation from the ice-surface in producing low temperatures. Fig. 94(b) should be compared with the average chart (Fig. 9(b)) which also shows a low over Antarctica. However the convolutions around the continent have been lost on the average chart.

Figure 94(a) shows a number of island stations in the Southern Ocean. At 60° S. the surface temperature is near 0° c. At Marion Island (47° S.) it is 10° c; at Macquarie Island (55° S.) it is 2° c and in southern Tasmania (43° S.) it is 9° c. Gough Island (not on the chart) at 40° S. was also recording 10° c. From 60° S. to 40° S. there is a general increase of about 1° c for each two degrees of latitude and Fig. 41 shows this rate of increase is continued to the equator.

Winds

Southern high latitudes record some very high wind-speed values. In sailing ship days the latitudes were named roaring forties, furious fifties and shrieking sixties. The antarctic coast, especially the eastern sector, is particularly windy. Cape Denison (142° E.), the base of Mawson's expedition, had a mean wind speed of 71 km/h and gales on 339 days in the year. Mawson named it 'the home of the blizzard'. Often stations along the coast have high mean values. Some of these gales are katabatic due to cold, dense air rushing down the smooth ice slopes under the influence of gravity. Such winds, depending on topography, are local. Sheltered places may be calm while nearby is a strong katabatic wind. Sometimes a sheltered bay or fiord may be calm while overhead a katabatic wind is blowing. This happens in Greenland as well as Antarctica. Other gales are due to the frontal funnel effect.

Figure 95 shows a typical situation in the north. A ridge of high pressure occurs over the inland ice and a depression near Cape Farewell is pushing warm air towards the cold air draining from the ice-cap. Winds accelerate in the narrowing gap between the front and the mountains (cf. winds through a gap in a mountain range on p. 56). Frontal funnel gales also occur on parts of the Icelandic coast and in many places in the antarctic. Flat-ice coasts on shelf ice in the antarctic, though apparently more exposed, are less liable to gales. Amundsen's base at the Bay of Whales had a smaller incidence of gales than Scott's base at McMurdo Sound even though the latter was 'sheltered' by higher ground behind.

Gales associated with deep depressions can be very dangerous to shipping in high latitudes. Strong winds whipping up high seas in sub-zero temperatures cause ice accumulation on decks, rigging and gear. The weight can prove fatal for a trawler. In February 1968 a Hull trawler was lost in such conditions while *sheltering* in a west Iceland fiord. Fig. 96 for 4 February 1968 shows a very deep depression centred just south of Iceland. At the

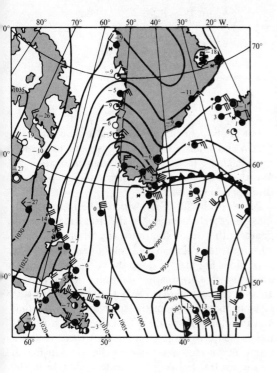

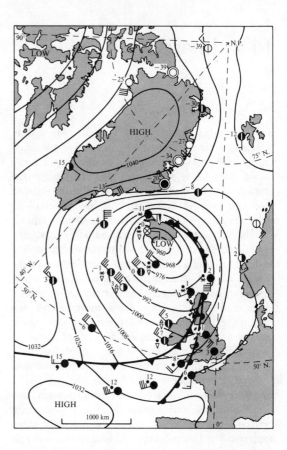

same time pressure was high over Greenland. Between the two is a strong gradient for easterly winds. Extreme north-west Iceland has a force 11 wind and a temperature of −11° C. Ocean weather ship *A* has a northerly force 9 and a temperature of −4° C. Commenting on the loss of the trawler, the third lost in a short period, a forecaster of the Icelandic meteorological service said, 'In many places the wind reached force 12 (100 km/h gusting to 150 km/h); there was heavy snow and the temperature was −8° C. Waves in the open sea were 13 metres high. The temperature was so low that even with a sea temperature of +4° C, you would be bound to get icing on a boat.'

Conditions such as these are a not infrequent feature of a high latitude marine climate. It is characterised by strong winds, high cloudiness and a high frequency of storms.

Precipitation

Ice-cap climates have a low amount of precipitation. Because of drifting, snow measurements are difficult but by striking the balance between loss and gain estimates are possible. Cold air can hold little moisture even when saturated (see the graph of Fig. 21) and anticyclones have subsiding air which is dry. The tundra extensions have only 250–300 mm water-equivalent but high latitude marine areas have more. Measurement at sea is difficult and islands are not entirely representative, but Jan Mayen, north of Iceland, has 340 mm, south-west Greenland 1100 mm, and the Aleutians 1350 mm. In the southern hemisphere the oceanic islands have values from 750 mm to 1600 mm. The precipitation in marine areas falls mainly as rain or wet snow.

11 Arid areas

Aridity implies lack of moisture but it is impossible to select one rainfall line that can serve as a boundary for arid areas. The 400 mm isohyet might serve in the Saharan region but colder areas with this amount of precipitation could have ample available moisture because evaporation is low. Rainfall in some seasons may be more effective than in others. On the southern edges of the Sahara the scanty rainfall comes in summer when temperatures are highest. On the Mediterranean border the equally scanty rainfall comes mainly in winter when temperatures are lower. Evaporation is therefore lower and the low rainfall is more useful for agriculture. 250–300 mm of rainfall in the higher latitude winter rainfall regime may be as useful as 500 mm in a lower latitude summer rainfall area.

Figure 110 shows the world distribution of arid areas. They include about one-third of the global land area. Most of the large deserts of the world are situated in the subtropical high pressure belts centred along 30° N. and S. On the poleward sides are transitional areas where distance from the ocean and shelter effects of mountain ranges give large regions with semi-arid conditions.

All deserts are similar in their aridity but they can be separated by temperature. Sometimes they are classified as hot or cold deserts. However, in summer so-called cold deserts may have very high temperatures (e.g. Ulan Bator). It is the winter temperature which is most distinctive and this can be used as the criterion. Thus there are:

(*a*) Hot deserts with no cold season. These are usually the trade-wind deserts.

(*b*) Mid-latitude deserts with a cold season. These are found mainly in Asia, but also in North America. Patagonia in South America has very similar characteristics.

Africa and the Near East
The chart of North Africa, Arabia and Iran for 1200 GMT, 14 July 1967 (Fig. 49), is fairly typical for the northern summer. Just after the solstice the noonday sun is overhead at about 20° N. and on this chart the inter-tropical convergence zone lies approximately along the same latitude following the axis of low pressure across Africa. Although it

is not easily defined over Yemen, it shows up over Oman and follows the trough eastward to the low over Pakistan (see monsoon trough in chapter 7). Weather south of the I.T.C.Z. on this chart was discussed in chapter 7.

North of the I.C.T.Z., which is here at about its most northerly position, is the arid area. On this chart, with a few exceptions, the whole area from Morocco to Pakistan has cloudless skies. The exceptions are mainly along the coasts where small amounts of fair weather cumulus are reported. Inland, convection over the hot land mixes the thin layer of moist air from the sea with drier air aloft and the clouds dissipate. The Libyan coast reports the presence of large cumulus and there is also a report of convection cloud over the Ahaggar massif in the central Sahara. A thundery shower in the mountains can result from such clouds. Away from the coasts, which on this chart show the effects of sea-breezes, almost everywhere the temperatures are in excess of 30° C. In Iraq, Baghdad reports 48° C. Over all this broad area blow the trade winds. Except in Syria where they have a westerly component caused by a trough of low pressure over Iran and Iraq, the winds are easterly or northerly. In the extreme west they are mainly easterly but over the eastern Mediterranean they are more northerly, an excellent example of the Etesian winds which blow with great regularity from mid-May to mid-October. The Etesian winds freshen by day and fall lighter at night. At several stations in Africa wind strength is sufficient to raise dust.

Figure 97 shows a chart near the spring equinox. The I.T.C.Z. lies roughly along 10° N. High temperatures occur just north of the convergence. Over the Sahara a small anticyclone occurs in the east and there is a depression south of the Atlas mountains bringing cold air in its rear. This depression is an example of cyclogenesis in the lee of the Atlas mountains in a deep west or north-west current (see Fig. 74). In the eastern desert the fall of temperature northwards from south Sudan to Egypt averages about 1° C per degree of latitude. From upper Volta to Algeria the temperature gradient is steeper and a cold front appears to be present. Algiers has rain. A high-level station in

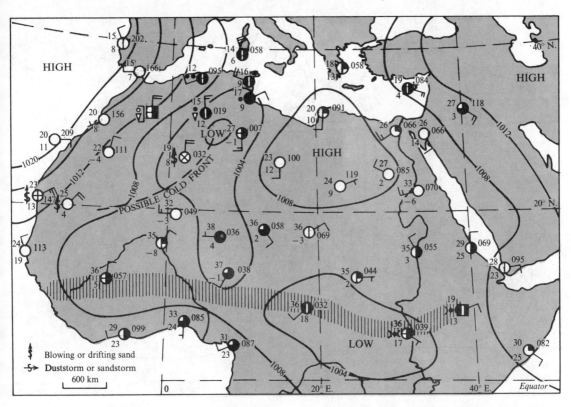

the Atlas mountains has a shower in sight and a shower is reported south of the Atlas. In the extreme west the north-east trade winds are strong enough to raise dust at 20° N. in Mauritania. There is also blown dust in the Tuat area to the west of the low pressure centre.

Figure 98(a) shows the surface chart 24 hours later. The small lee depression has moved north-east and deepened considerably, bringing cold air well south over the Sahara in its rear. Gao on the upper Niger shows a temperature drop of 5° C between the 13th and 14th and along 20° N. the drop is 9° C. The cold front has reached the I.T.C.Z. Well ahead of the low pressure system, over the Sudan and Egypt, temperatures have changed little between the two charts. In the rear winds in the cold air are sufficiently strong to give widespread blown dust and dust-storms. Ahead of the depression the southerlies have strengthened over Libya giving much rising dust. Warm southerly winds ahead of depressions are known as sirocco over most of the Mediterranean and as khamsin in Egypt. These siroccos vary in character between Africa and the Mediterranean islands and peninsulas. In North Africa the sirocco is very hot in summer and very warm in winter but always dry. Very low humidities

of less than 10% have been recorded. Fig. 97 shows the low dewpoints (−3 to −9) of the air over the desert which will be drawn into the southerly stream. Siroccos also carry dust up to quite high levels. Crossing the relatively cool water of the Mediterranean the warm sirocco air is cooled from below: a low-level inversion forms with a layer of cloud beneath it. Drizzle may fall from this stratus if the sea track is sufficiently long. The air is still hot and being humidified also becomes 'sticky'. As the front of the depression approaches, the rain which falls may be carrying dust. 'Blood rain' is not uncommon in southern Italy and 'red snow' falls on the mountains in winter. From a well-formed depression it is estimated that 1 to 2 million tons of African dust may fall on Europe. In the western Mediterranean fifty siroccos may occur each year but in the east strong southerly winds from the desert are less frequent. The average number of khamsins in Egypt is about fifteen to twenty. Spring is the main season. The frequency at Alexandria is (ten year period):

J.	F.	M.	A.	M.	J.	
Jy.	A.	S.	O.	N.	D.	Total
6	6	40	50	46	16	
0	6	14	6	6	6	202

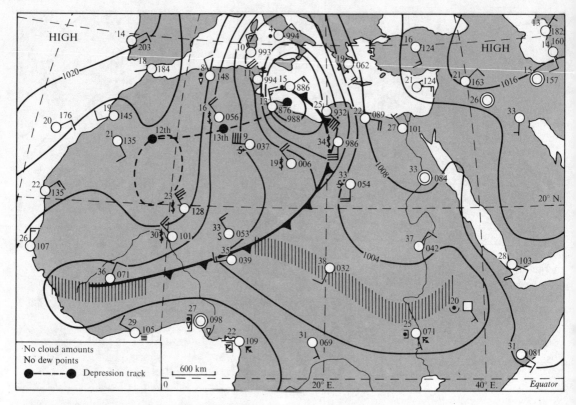

There are very many local winds in the Mediterranean area. One book lists forty-two special names. Most important are the sirocco type (warm, southerly and dusty) and the mistral–bora type (cold northerly or easterly) (see chapter 9). As a depression travels along the Mediterranean the warm dusty winds will start as sahat in Morocco, become siroccos in the central area and khamsin in Egypt. Progressively from west to east they are brought to an abrupt termination by the arrival of cold air. On Fig. 98(a) notice the large differences in temperatures across the cold front. Behind the front bora-type winds will emerge through gaps in the northern mountain wall into the Mediterranean. Over the sea cumulus will build up as the cold air is heated from below by relatively warm sea. The clouds bring heavy showers in the southern Mediterranean and on the African coast (cf. Algiers on Fig. 98(a)).

Further east Fig. 98(a) shows the anticyclone continuing over the Caspian Sea region. It is instructive to compare the afternoon temperatures in Iran and Iraq in mid-March (near the equinox – Fig. 98(a)) and in mid-July (after the solstice – Fig. 49). Individual stations show differences of over 20° C.

Dust in the atmosphere

As an example of weather in arid lands the chart for 14 March 1962 (Fig. 98) is most noteworthy for the widespread dust-storms already mentioned. Dust is an important element in the climate of all desert and desert-fringe areas. Strong winds blow the loose dust and with rising air in frontal situations it may reach considerable heights. In more settled conditions intense heating of the ground during the day sets up a steep lapse rate favouring ascent of air and gusty winds at the surface. Wherever there is light dusty soil it is caught up in the ascending air to give a sandstorm – a small whirling column of dust of short duration. These are called dust-devils or, if of greater intensity, simooms. They occur most often during the hottest part of the day and the year. At night the dust settles out of the lower layers but persists aloft and eventually the whole air mass may become hazy from the dust.

Dust from the desert is carried in the harmattan air over the savana and equatorial forest regions. Harmattan haze gives poor visibility at Lagos airfield and Saharan haze can be carried on easterly winds as far as the Azores 800 miles out in the Atlantic. Occasionally Saharan dust reaches England and Wales, as on 30 June 1968 when it is estimated that about one million tons of dust fell over England and Wales south of a line from Preston to Scarborough. Quartz was the major constituent but there was some clay, and also freshwater diatoms. The latter suggests an origin in the periodically flooded upper Niger district, with the major constituents coming from the southern edge of the Sahara near the Ahaggar mountains.

In the Sudan special summer dust-storms associated with southerly winds and thunderstorms are called *haboobs*. The arrival of a haboob is heralded by the approach of a cumulonimbus cloud from the south. Rain streaks falling from the cloud evaporate in the dry air. The cloud is accompanied by columns of dust which at a distance look like smoke from a line of fires. Gradually the columns merge to give an advacing wall of dust which rises up to the cloud base. At Khartoum the dust is often reddish and arrives with an increase of wind. Temperature drops 2° to 3° C with the arrival of the dust, which penetrates inside houses covering everything with a layer of dust. It is a weather phenomenon with a high nuisance value. At Khartoum haboobs occur usually in the afternoon and with increasing frequency in May, June and July. Seventeen is the average annual number.

Temperature range

The subtropical anticyclones over the continents provide the areas with the highest recorded temperatures. Invasions of cold air come not only from the north but also from the south – the air from equatorial regions has lower temperatures. Just as incoming radiation is little impeded by the dry cloudless atmosphere giving high day temperatures, there is also a high value for outgoing radiation and at night surface temperatures fall considerably. Ground frost is not uncommon and screen temperatures may fall below freezing point. The graph for Fort Flatters (Fig. 98(b)) shows the great daily and annual range. Below freezing temperatures have occurred at El Obied (13° N. and 550 m), and at Wadi Halfa (21° N. and 120 m) there are many records of temperatures several degrees below freezing.

Cold-water coasts

The chart for 14 July 1967 (Fig. 49) shows a steep gradient of temperature from the coast inland in north-west Africa. Port Etienne records 24° C. Inland 37° C is reported. Further north contrasts

are even more marked. On the March charts (Figs. 97 and 98(a)) Port Etienne has a similar temperature to its July figure: 22° C. The contrast with inland stations is much less than in July. In both March and July dewpoint at Port Etienne and along the coast generally are high as compared with inland stations. This region is a good example of a cold-water coast.

The main aridity control is the dry subsiding air in the eastern end of the subtropical anticyclone. Winds blowing around the high pressure cell blow parallel to the shorelines or marginally off-shore. The aridity is reinforced by the cold coastal currents which retard evaporation. *Upwelling* water further reinforces the coldness of the current. Subsidence in the high pressure cell creates a temperature inversion not far above the surface. However, the wind strength usually forces a normal lapse rate in the lower layers so the inversion varies from the surface up to 1000 m. Often a layer of stratus cloud forms underneath the inversion.

The weather chart of South America for July 1957 (Fig. 47(a)) shows a complete cloud cover with a temperature of 16° C at Iquique (20° S.) and at 4° S. in Peru the temperature is only 21° C. Conditions vary very little on Fig. 47(b) four days later. Here is a very great length of arid coast. Antofagasto on the southern tropic has 13 mm rainfall annually: Lima, 1600 km to the north, 40 mm. The aridity of the coast is enhanced by the close proximity of the Andes which act as a geographical control on the position of the subtropical anticyclone. The Andes also give shelter from the easterlies. Another factor which is accounted important in determining the length of the arid zone is the consonance of the curve of the coast with the anticyclone curvature of the South Pacific cell. Where the coast turns east at about 5° S. coastal rainfall increases rapidly and Colombia registers some of the highest rainfall totals near sea level in the world. The station at Andagoya, 5° N., is 62 m above sea level and records an average rainfall of 5378 mm.

In North America the equivalent coast shows less extreme aridity over a shorter distance. Fig. 77 drawn for local time of 1500 hours at 120° W. shows a large difference between coastal and inland temperatures. The northern tip of Vancouver Island and San Francisco 1500 km south record the same temperature. Between the two a much lower temperature occurs in fog off the north California coast. Such fog banks often extend across the entrance to San Francisco Bay. Fig. 59 shows stratus off the Californian coast.

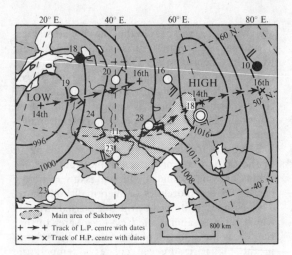

Along the Namib, southern Africa has a similar dry littoral strip. The west Australian counterpart north of Perth is less arid by comparison. Anticyclones move more quickly eastwards across Australia, rather than remaining stationary, and this allows air with varying degrees of stability to affect the western coast bringing more rain. The driest parts have about 250 mm annually, which is twice as much as lower California and at least five times as much as the three other cold-water coasts.

Interior mid-latitude conditions

The development of anticyclonic conditions over the continental interiors in spring and summer can give rise to very hot dry spells of weather called sukhovey (literally dry winds) in the U.S.S.R., where great damage may be done to crops in the Ukraine and southern Volga Basin.

Figure 99 shows a typical sukhovey situation. The anticyclone centred north of the Caspian Sea on 14 August had moved from a position north of the Black Sea on the 11th. In this position a dry northerly air stream was drawn down the Volga river region and relative humidity had fallen to 20% by the 12th. As the anticyclonic centre moved eastwards the winds over the Caspian lowlands and eastern Ukraine became south-east or south. Fig. 99 represents 0900 local time for the main sukhovey area. Already a temperature of 28° C is shown. By afternoon temperatures reached the upper thirties.

The high temperatures and low humidities are the result of two processes. Firstly the air in slow-moving anticyclones is warmed adiabatically by subsidence and its relative humidity decreases. Secondly intense surface heating during the day under clear sky conditions causes dry adiabatic or even super-adiabatic lapse rates to develop in the layer of air below the anticyclonic inversion. Thus strong convection currents develop in this layer and in the descending currents very dry air from aloft is brought to the surface. As this hot dry air causes transpiration from plants to exceed the absorption of soil water by the roots, considerable damage to plants results. Leaves wilt, go yellow and eventually wither. Temperatures may reach 40° C with relative humidities below 30%. Severe damage occurs in a very short time.

Figure 99 indicates how this spell of sukhovey ended. The anticyclone continued to move eastwards into Siberia and the low pressure trough over the Baltic Sea on the 14th moved to a position near the Volga river by the 16th. Although temperatures remained high (36° C) the westerly air flow brought higher humidities and an end to the dessicating conditions.

The part of the Soviet Union most affected by sukhovey is shown on the map. Only rarely does it occur simultaneously over the whole area. Similar, but less extreme, conditions may occur elsewhere even at the forest edges further north. Extensive fires in the forests and peaty ground just east of Moscow in August 1972 were helped by exceptional dry spells.

The interior mid-latitude areas in the U.S.A. experience similar situations but on a less extensive and extreme scale due to the smaller land mass.

12 Mountain areas

In the foregoing chapters it has been shown how the presence of mountain ranges and upland areas affects the flow of air in the troposphere. In the westerly wind belt of mid-latitudes these areas are responsible for the initial development of the Rossby waves, upper warm ridges and cold troughs which have such a marked influence on surface weather conditions and so on the climate of large parts of the earth's surface.

Within mountain regions the effects of altitude, aspect and orientation in relation to the general wind systems are responsible for marked modifications in the climatic elements in comparison with those experienced at low altitude away from the mountains.

The decrease of pressure with height from a mean sea-level pressure of 1013 mbar to 500 mbar at 5–6 km and to 300 mbar at c. 9 km ($\simeq$ the summit of Mt Everest), together with the decrease in absolute humidity because of low temperatures, and in the quantity of dust particles, means that a higher proportion of the short-wave and very short-wave (ultraviolet) radiation from the sun penetrates to ground level. This results in very high ground surface temperatures being experienced on sunward-facing slopes. Thus in the Ötztal in the Austrian Alps a surface temperature of 80° C was recorded on 7 July 1957 on a 35° south-west slope at an altitude of 2070 m, whereas the surface temperature on a north-east slope at the same time was only 23° C.* Because of the low density of the air at high altitudes air temperature decreases rapidly from the ground upwards.

Under partly cloudy conditions surface temperatures at high altitudes fluctuate rapidly between the sunny and cloud–shadow periods and variations of the order of 10° C have been recorded within a few seconds.

Although a high proportion of possible insolation is received by surfaces able to receive it, the greatly dissected character of mountainous areas means that many slopes are often in the shadow of neighbouring mountains and so only receive a proportion

*R. Geiger, *The climate near the ground* (Harvard and Oxford, 1966).

of the possible insolation under clear sky conditions. This is particularly so in extra-tropical areas, especially in winter. Thus at Obergurgl (46° 52′ N. 11° 2′ E. alt. 1940 m) in a southern tributary valley of the Inn, where the ridges to the west, south and east frequently rise to over 3000 m, the restriction of the horizon causes a loss of 10% of the global radiation under cloudless conditions in June and 60% in December.

On 17 April 1966 Durham University students made a comparative study of the air temperatures on the south- and north-facing slopes of a west to east section of the Lima valley (44° 01′ N. 10° 37′ E.) in the Tuscan hills of Italy. Synchronised temperature readings were taken at 1.5 m above the ground at intervals of 10 minutes between 1130 and 1630 hours at altitudes of 300 m and 500 m on both sides of the valley. The conditions were not ideal for the development of strong thermal contrasts between the slopes since cumulus cloud cover varied between 4 and 7 octas and a south-west wind gave speeds of 4–12 km/h at the low level stations and 30–40 km/h at the higher ones and temperatures fluctuated rapidly as a result. The results obtained are shown in Table 12.1. The altitude of the mid-day sun was 53°.

Table 12.1 *Comparison of air temperatures on the south- and north-facing slopes of a west–east section of the Lima valley*

	South-facing slope		North-facing slope	
Angle of insolation at mid-day	86°		20°	
Height	300 m	500 m	300 m	500 m
Maximum temperature	20.0°C	22.8°C	17.2°C	15.0°C
Mean temperature	16.6°C	18.9°C	15.5°C	13.3°C
Minimum temperature	13.9°C	15.0°C	14.4°C	12.8°C
Max. fluctuation in 10 minutes	4.4°C	5.0°C	3.3°C	1.7°C

That insolation is an important factor in the local land use is reflected in the fact that on the south-facing slope vines, cereals and vegetables were grown up to 400 m with maquis above, whilst the north-facing slope was forested.

Night-time radiational cooling

The occurrence of low temperatures on valley floors and upland basins on clear, calm nights as a result of the loss of heat by outgoing radiation occurs particularly in upland areas.

For 103 days between 30 January and 13 May 1965 thermographic records were made in the Cold Canyon, Santa Monica Mountains, California (Fig. 100(*a*)). During the period there were no major intrusions of polar air. Fig. 100(*b*) shows the thermographic records for the four stations from 15 to 20 February 1965 when conditions were particularly favourable for night-time radiation. Table 12.2 shows a selection of the results obtained during the whole period of study.

The study over the whole period shows that:

(*a*) The climate is much more severe at the two lower sites than on the hillside or mountain peak.

(*b*) The average minimum temperature was 7°C lower at the lowest site than at the mountain peak, which is 5 km away and 596 m higher.

(*c*) The frost pocket at Berrian Ranch experienced an extreme minimum temperature 8°C lower than the hillside station, Stunt's Ranch, 91 m higher.

During the period 15–20 February the clear-sky, anticyclonic conditions favoured day-time insolation as well as night-time outgoing radiation. The maximum day-time temperatures were virtually the same except at Topanga Lookout where they were 3° to 4°C lower (Fig. 100(*b*)). Night-time minima, however, fell to very much lower levels at the two

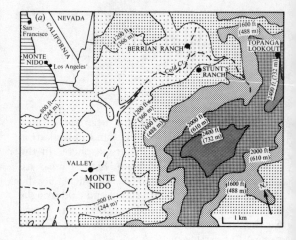

lowest stations, the lowest temperature of −6.6°C being recorded on two successive nights in the Monte Nido valley. The diurnal range at the lowest stations was consequently very great, exceeding 30°C in the Monte Nido frost pocket on four successive days, when on 18 February it rose from −4°C at 6.50 a.m. to 29.2°C at 1.30 p.m.

The night-time minima show that from 16 to 20 February the inversion of temperature between the lowest and highest stations was 18°C – a 1°C increase in temperature for each 54 m increase in altitude. Under cloudy conditions the night-time inversions are very much weaker.

The evidence collected in the study indicated that the very low temperatures in the frost pockets were not due to the down-hill flow of chilled air (katabatic winds) but to radiational cooling, since the air in the bottom of the hollows was cooler than that above from the time of the onset of the nocturnal inversion.

Table 21.2   *Thermographic records made in Cold Canyon from 30 January to 13 May 1965*

Station	Altitude (metres)	No. of days out of 103 when minimum temperature was: 0°C	10°C	Minimum temperature Average (°C)	Extreme (°C)	No. of days out of 103 when temperature range was: 11°C	22°C	Ranges Mean (°C)	Extreme (°C)
Topanga Ranch (peak)	753	0	26	8.1	0.3	86	0	8.3	16.6
Stunt's Ranch (hillside)	395	0	22	7.9	2.4	35	0	12.2	21.1
Berrian Ranch	304	32	0	1.3	−6.3	11	32	18.3	28.8
Monte Nido Valley (pocket)	159	37	1	0.9	−8.8	9	32	18.8	33.3

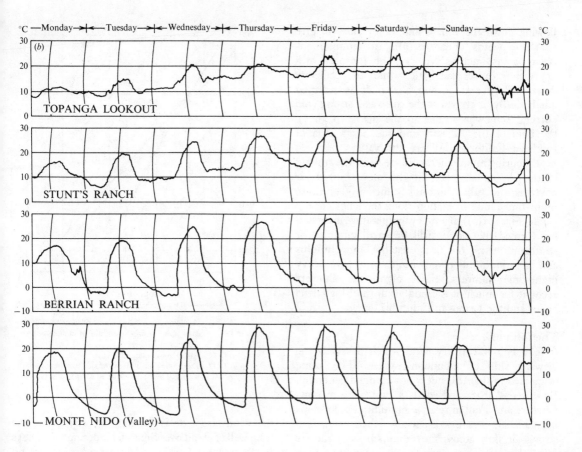

°C —Monday—→|←—Tuesday—→|←—Wednesday—→|←—Thursday—→|←—Friday—→|←—Saturday—→|←—Sunday—→|←— °C

TOPANGA LOOKOUT

STUNT'S RANCH

BERRIAN RANCH

MONTE NIDO (Valley)

Mountain winds

Under clear-sky, calm-air conditions the greater
intensity of insolation on the higher mountain
slopes than in the valley bottoms results in them
becoming high level sources over which the air
expands and rises to be replaced by an up-slope flow
of air from the valley bottoms. Such winds are called
anabatic (Fig. 101(*a*)) and the flow commences in the
early morning and is often followed by a flow up the
valleys, even against the regional pressure gradient.
Cumulus clouds may build up over the mountains
during the morning, developing into cumulonimbus
clouds, which give heavy showers, during the afternoon. Over the Cordillera of the north Andes the
rainfall in the rainy season is mainly the result of
such day-time convectional activity.

Night-time radiational cooling of the ground
under clear-sky, calm-air conditions is often, but
not always, greater on the higher mountain slopes
than in the valley bottoms, especially if the higher
ground is snow covered. The increasing density of
the chilled air at higher levels then results in it

101(*a*) Anabatic winds in daytime

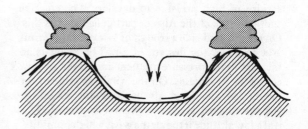

101(*b*) Katabatic winds at night

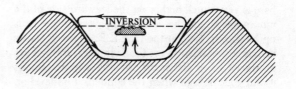

flowing down the valleys and valley sides to give *katabatic* winds (Fig. 101(*b*)). In mountainous coastal areas, such as Norway, the approach of a depression causes katabatic winds to accelerate down valleys and, although the air is warmed adiabatically, it arrives at the coast as a strong cold current. Somewhat similar are the mistral (p. 56–7), the bora of the Adriatic and the Santa Ana of southern California. Where the valleys between neighbouring mountain ranges are wide, the sinking of the chilled air may lift the more humid air covering the valley bottoms sufficiently to cause night-time cloud formation along the length of the valley (Fig. 101(*b*)). If a valley temperature inversion has formed the clouds will be small cumulus or stratocumulus; if the air is unstable large cumulus and cumulonimbus clouds build up to give heavy night-time showers. Such night-time showers account for much of the rainfall in the Magdalena valley and the Llanos of Columbia.

The föhn effect

A föhn is a warm, dry wind which descends the leeward side of mountain ranges. The name originated in the Alps but is now applied generally. Its onset is characterised by a rapid rise in temperature and a fall in relative humidity. It develops in stable air streams, especially when there is a strong air flow across the mountains and the air mass approaching the windward slopes has a high humidity.

Under such conditions the deflection and ascent of air on the windward side of the mountains causes a ridge of high pressure to develop. The concave southern side of the Alps in particular favours this. On the leeward side a trough of low pressure forms as a result of the descent of air. The result is an increase in the pressure gradient and in the air flow across the mountains. On the windward side of the mountains ascending air cools at the D.A.L.R. (10° c/km) until condensation level is reached. This is at a low altitude if the air mass has a high humidity. Thick layers of cloud form above condensation level, above which the saturated air cools at the S.A.L.R. (6° c/km). Heavy orographic precipitation often occurs to cause a marked decrease in the absolute humidity of the air. Because of this cloud base is very much higher on the leeward side of the mountains where the descending air mainly warms up by compression at the D.A.L.R. Fig. 102 shows in diagrammatic form the föhn temperature changes produced in an airstream crossing a mountainous island 3000 m high when the condensation level on

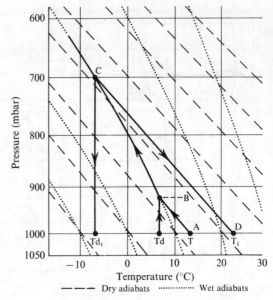

T	Air temp. on windward side of mountains at A
Td	Dew point on windward side of mountains at A
B	Condensation level
B to C	Rising air cooling at SALR
C to D	Descending air on leeward side of mountains warming by compression at DALR to attain temp. T_1 at D
Td_1	Dew point of air at temp. T_1

the windward side is 500 m. Fig. 103 shows the föhn wall of cloud over Signy Island, South Orkneys (61° S. 43° W.). On the windward side of the island orographic layer clouds formed in the moist surface and upper layers. The wave form of the upper layer of clouds indicates the stability of the air-mass. The evaporation of the lower clouds on the leeward side of the mountains is very evident.

Figure 104(*a*) shows the synoptic situation when a pronounced föhn was blowing in some localities on the northern side of the Swiss Alps and Fig. 105 the autographic records at Altdorf before and during the föhn period which began at 1000 hours on 15 February 1967.

The general situation on this occasion showed that a blocking anticyclone over eastern Europe was slowly retreating as the frontal systems of a deep depression south of Iceland edged eastwards across north-west Europe and the western Mediterranean where cyclogenesis occurred (see pp. 58–59). Ahead of the fronts the pressure gradient steepened and south-westerly winds at all heights strengthened.

As a result of the north to south air flow over the mountains a ridge of high pressure, termed a föhn nose, formed over the Alps and the Lombardy plain with a trough of low pressure over the Alpine

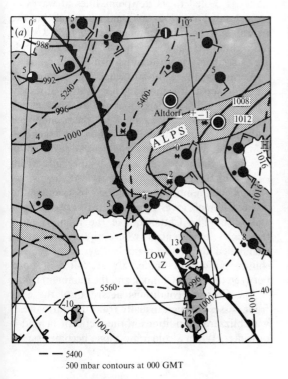

5400
500 mbar contours at 000 GMT

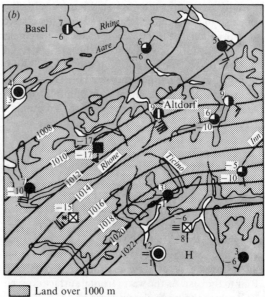

Land over 1000 m

foreland. The pressure gradient became particularly steep over the westernAlps which also retarded the eastward movement of the fronts.

Cloudy conditions prevailed in the moist southerly air stream. The cloud types being reported were stratus, stratocumulus and altostratus over and to the south of the Alps. Over the Lombardy plain cloud base was 300–1000 m and the temperatures and dewpoints were at or near 0° C.

North of the Alps cloud bases of 3000–5000 m were being generally reported and dewpoints were as much as 12° C below the dry bulb temperatures. The autographic wind records for Altdorf showed that between 0700 hours on 13 February 1967 and 1000 hours on 15 February 1967 calm conditions prevailed with diurnal variations in temperature and relative humidity related to day-time heating and night-time cooling. At 1000 hours on 15 February the föhn began to blow with the northerly wind increasing to 40 km/h by 1200 hours. At 1215 hours the wind suddenly changed direction to south with a mean speed of 55 km/h and rapidly alternating gusts and lulls between 90 and 20 km/h. This persisted until the morning of 17 February. Between 0900 hours and 1300 hours on the 15th the temperature rose from −7° C to 10° C and relative humidity fell from 95% to 15% (Fig. 105). Over the next 48 hours the diurnal temperature range was only 3–4° C. The relative humidity remained at 15% until mid-day of 16 February. Fig. 104(b) shows the synoptic situation at this time with the föhn also blowing in the Rhone valley. The relative humidity then rose to 25% by 1800 hours, to remain at this level until 1000 hours on the 17th. A rapid increase to 95% at 0400 hours on the 18th followed, by which time the temperature had fallen to −2.5° C. As the records indicate, there were minor returns to föhn conditions on the afternoon of 18 and evening of 19 February.

The Chinook

This is the North American equivalent of the föhn (p. 90). It occurs when there is a stable westerly air flow across the Rockies. The onset is associated with the formation of waves on, and the breakdown of, Pacific warm fronts east of the Cordillera. Fig. 106 shows the synoptic situation for one such occurence. South of 53° N. a Pacific warm front edged slowly eastwards against the dense, very cold cP air to the east. As the warm frontal surface crossed the Rockies the wave shown at 50° N. 110° W. and the lee depression centred near Calgary formed. The absence of precipitation along the warm front is characteristic of this situation. A rapid rise in temperature took place as the warm sector air descended the eastern slopes of the Rockies.

On 9 December Calgary, like the rest of the continental interior, was in the cP air mass up to 2200 hours. Up to that time winds were light southerly and the maximum temperature was −17° C at 1500 hours. Between 2200 and 2300 hours the wind fluctuated before becoming a steady westerly. The temperature rose from −22° C to 0.5° C between 2300 hours and 0900 hours on 10 December and remained above freezing point for the next 23 hours. During this period the Chinook wind was blowing from the west at speeds of up to 50 km/h. Between 0800 and 0900 hours on 11 December the wind backed to south-west; the temperature fell to −2.5° C by 1000 hours and the Chinook was coming to an end. The upper air descent for Edmonton showed a marked temperature inversion of 18° C in the cP air below 895 mbar, a frontal mixing zone between 895 and 850 mbar, and warmer, stable air above 850 mbar.

The descent of the air is caused by the flow of air from the higher pressure on the windward side of the mountains to the low pressure on the leeward side and by the generation of a wave flow in the lower layers of the stable air by the high ground (Fig. 107). The wave flow extends down wind over the lower ground. When the air rising into the crests of the waves is cooled to its dewpoint lines of waves or arch clouds form (Fig. 108) and run parallel to the mountain front. The 'Chinook Arch' of the Rockies has a cloud base at about 3000 m. The cloud bands are stationary, with condensation continuously occurring on the windward edge and evaporation on the leeward. It is the wave flow which causes the descent of air from heights above the crest of the mountain ranges and accounts for the pronounced sudden rises in temperature when the 'shooting flow' displaces the arctic air at ground level to the east of the Rockies. Rises of temperature from −35° C to 10° C accompanied by strong, dry winds are not uncommon in midwinter. In summer the warm Chinook gives temperatures between 30° C and 37° C. The Chinook, like the föhn, is most frequently experienced where the warm air is channelled down the major valleys. As Fig. 107 shows, pockets of colder air are often trapped between the warm Chinook and the mountain front.

The Chinook conditions persist until the cold front passes through, usually from the north-west, with blizzards conditions in the arctic air behind it in winter. When a cold anticyclone becomes estab-

105 Autographic record of temperature and relative humidity, Altdorf, Switzerland, 13–19 February 1967

106 (*centre*) Synoptic chart, 0000 GMT, 11 December 1966

107 (*below*) Chinook flow over the prairies

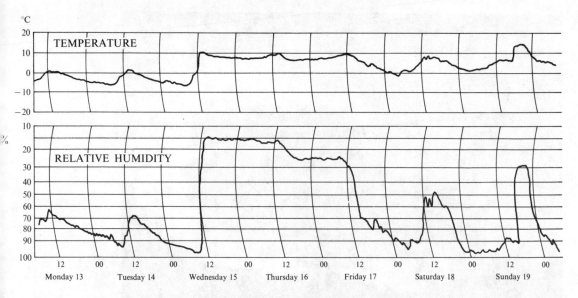

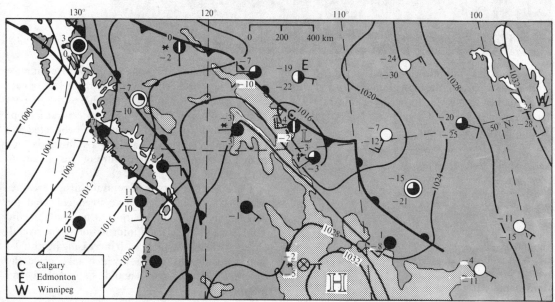

C Calgary
E Edmonton
W Winnipeg

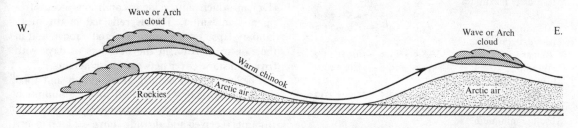

lished in the Chinook belt of Alberta the frontal systems and associated Chinook conditions are experienced much further south in the United States.

The relative mildness of the winters in the westernmost parts of the high plains compared with areas further east is illustrated by Table 12.3, the temperature records of Calgary (51° N., 106 m) and Winnipeg (50° N., 234 m).

Table 12.3 *Temperature records of Calgary and Winnipeg*

	N.	D.	J.	F.
	°C	°C	°C	°C
Absolute maximum				
Calgary	21.5	19.5	16.0	19.0
Winnipeg	21.5	11.0	8.0	8.5
Mean monthly maximum				
Calgary	17.5	11.0	10.5	12.0
Winnipeg	10.5	2.0	−1.0	2.0
Mean daily maximum				
Calgary	3.5	−1.5	−4.5	−2.0
Winnipeg	−1.0	−9.5	−14.0	−11.5
Mean daily minimum				
Calgary	−8.5	−13.5	−16.5	−14.5
Winnipeg	−10.0	−19.0	−25.0	−23.0
Mean monthly minimum				
Calgary	−23.0	−28.5	−33.0	−31.0
Winnipeg	−24.5	−34.0	−39.0	−36.0
Absolute minimum				
Calgary	−35.0	−43.0	−44.5	−45.0
Winnipeg	−36.5	−49.0	−44.0	−44.0

The detailed study of mountain climates in most parts of the world is handicapped by the lack of data on the various climatic elements due to the sparsity of meteorological stations. Only in central Europe and in North America has data been collected at a large number of stations and for a sufficiently long period of time for a true picture of the climate to emerge.

The Alpine countries of central Europe are ones in which sufficient data have been collected for detailed statistical information to be available and Table 12.4 gives selected information of the overall changes which take place with increasing height. To fully visualise the changes which take place the different sets of data should be graphed and interpreted.

In interpreting the data it should be borne in mind that the area experiences a continental climate (Dc) in which anticyclonic conditions occur frequently in winter. This is reflected in the mean January lapse rate between 800 and 1400 m and in the rate of increase in the numbers of days with frost and snow cover between these heights.

The increase of annual precipitation and total depth of new snow might lead to the assumption that there is always a *direct* relationship between the amounts received and altitude above sea level in any given mountainous area. A careful comparison of detailed mean annual rainfall and topographic maps of a mountainous area will show that factors other than altitude, such as orientation in relation to the prevailing rain-bearing winds, influence the mean rainfall over a long period of time. Thus in the Snowdonian mountains of North Wales records show that the same annual rainfall can be received by places which differ as much as 500 m in height.

Above 1500 to 2000 m the low moisture content of the air due to the low temperatures (Fig. 21) is reflected in a decrease in precipitation with altitude. This is well illustrated in the case of Kilimanjaro (Fig. 109), where the annual precipitation is greatest (3000 mm+) on the southern slopes of the mountain at about 2150 m, decreasing to less than 500 mm at the summit. Most of the rainfall on the southern side occurs in March, April and May during the northward passage of the inter-tropical convergence zone and is mainly orographic in character, since an inversion at 4000–5000 m forms a ceiling to convection. The deflection of the south-easterly monsoon winds behind the I.T.C.Z. accounts for the westward extension of the rainy area on the southern flanks and the northward extension on the eastern flanks of the mountain. Under these quasi-stable conditions the northern and western sides of the mountain are in a rain shadow. The northern side of the mountain receives most of its rainfall during the southward passage of the inter-tropical con-

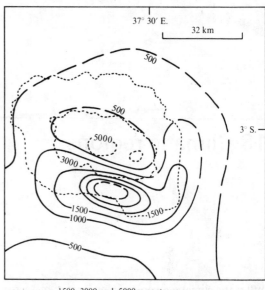

-------- 1500, 3000 and 5000 m contours
———— Suggested isohyets in areas with no reporting stations

vergence zone in November and December, heralding the onset of the north-east monsoon which lasts until March. During this period the high-level inversion frequently breaks down and convectional showers make an important contribution to the October to March rainfall.

Table 12.4 *Changes in climatic conditions with height above sea level in the eastern Alps*

Height in metres	Mean air temperature		Annual number of days with:				Annual precipitation	Total depth of new snow
	Jan.	July	Frost	Snow cover	Snowfall	Dry ground		
	°C	°C					mm	cm
200	−1.4	19.5	93	38	27	187	615	51
400	−2.5	18.3	98	55	32	173	750	116
600	−3.5	17.1	115	81	38	160	885	182
800	−3.9	16.0	131	109	45	147	1025	247
1000	−3.9	14.8	139	127	53	133	1160	313
1200	−3.9	13.6	147	138	62	120	1295	379
1400	−4.1	12.4	154	152	73	107	1430	444
1600	−4.9	11.2	162	169	85	93	1570	510
1800	−6.1	9.9	175	189	98	80	1700	575
2000	−7.1	8.7	187	212	113	67	1835	641
2200	−8.2	7.2	202	239	128	53	1970	707
2400	−9.1	5.9	219	270	143	40	—	—
2600	−10.3	4.3	240	301	158	27	—	—
2800	−11.3	3.2	264	332	173	13	—	—
3000	−12.4	1.8	294	354	188	0	—	—

Source: R. Geiger, *The climate near the ground* (Harvard and Oxford, 1966).

Part 3
Climatology

13 Climatic regions

When the climatic records of a number of meteorological stations within a small area of the earth's surface are studied for a single year or, as averages, over a period of 30 years, it is appearent that each place possesses unique characteristics but, at the same time, has many similarities with its neighbours. A comparative study of the climate records for one part of the earth's surface with those for other areas in different parts of the same or other continents reveal close similarities in some cases and marked differences in others. The recognition of such similarities and differences has prompted attempts to identify different types of climate, to classify them according to their particular characteristics and map their spatial extent.

One of the earliest classifications was the division of the earth into tropical, temperate and polar climates, using the tropics, $23\frac{1}{2}°$ N. and S. and the arctic ($66\frac{1}{2}°$ N.) and antarctic ($66\frac{1}{2}°$ S.) circles as dividing lines, with temperature conditions being the only ones considered as having any importance.

Later the realisation that, although temperatures do decrease with increasing latitude, they do not necessarily do so, led to the drawing of selected isotherms to delimit temperature zones. Supan, a German climatologist, in 1879 suggested the mean annual isotherm of 20° C, which approximates to the polar limit of palms, as the polar boundary of tropical climates, and the 10° C isotherm for the warmest month, which approximates to the polar boundary of tree-growth, as the division between temperate and cold climates.

Köppen, another German climatologist, developed Supan's ideas and in 1884 suggested the division of the earth into the following temperature zones:

Tropical	All months over 20°C
Subtropical	4 to 11 months over 20°C
	1 to 8 months between 10°C and 20°C
Temperate	4 to 12 months between 10°C and 20°C
Cold	1 to 4 months between 10°C and 20°C
	8 to 11 months below 10°C
Polar	All months below 10°C

The temperature zones were then subdivided on the basis of seasonal incidence and amounts of annual rainfall.

Between 1900 and 1931 Köppen produced a number of modifications of his classification and subsequent modifications have been made by other climatologists especially in Germany and the U.S.A. The modified Köppen classifications are to be found in many texts of physical geography, especially those published in continental Europe and the U.S.A.

G. T. Trewartha is one of the leading American climatologists who has produced a number of modifications of the Köppen classification since the mid-1930s. In the latest edition of his book *An Introduction to weather climate* (McGraw-Hill, 1968) his latest revision differs so much from the Köppen classification that, whilst acknowledging his debt to the German climatologist, it is no longer attributed to him.

Trewartha, like Köppen in his later classifications, chooses the 18° C isotherm for the coolest month as the polar limit of tropical climates, since many tropical plants do not thrive beyond this limit. He confines it, however, to coastal areas and in inland areas takes the equatorial limit of frost as his boundary. For extra-tropical regions the number of months with temperatures above 10° C, considered by many as the lowest temperature for human comfort and as a significant threshold for the onset of active growth in the case of plants, is important. Climates with fewer than 4 months with an average temperature of 10° C are of little use agriculturally. As a boundary the isopleth of 8 months with a temperature of 10° C also approximately coincides with that forming the polar limit of many subtropical plants and, on the western side of the continents, with the boundary between those areas experiencing summer drought and those with no marked dry season at this time of the year.

Because of its relative simplicity and usefulness Trewartha's classification is used in this book, and Fig. 110 shows his climatic regions for the world and Table 13.1 summarises the main characteristics of each.

The boundaries of the climatic regions in Fig. 110 are based on the average of the climatic conditions over a 30-year period. Using the same criteria, however, the climate of stations can be classified on a yearly basis and when this is done over a 30-year (or longer) period many stations are found to experience one type of climate in some years and a different type (or types) in others.

In 1954 S. Gregory used the modification of Trewartha's 1943 version of Köppen's classification shown in Table 13.2 to produce a map (Fig. 111) of the climatic regions of Europe for the period 1901–30 based on climatic year analysis.

The climatic data which were analysed to produce the map were from stations below 950 m which had reliable records for most or all of the period 1871–1940. The main climatic boundary between two particular types of climate was drawn where 15 of the 30 years fell into each category. The core areas of each climatic type are those which experience that type for at least 25 of the 30 years. Areas with 15 to 25 years in one category are designated transition areas and their width is a true indication of the rapidity of horizontal change in climatic conditions. The ease with which air masses from different source regions can cross the lowlands of central Europe is reflected in the pronounced width of the transition zone between the areas experiencing predominantly maritime or continental climates. The significance of mountainous areas, such as those in Scandinavia and the Balkans, in stabilising boundaries and restricting the width of the transition

Table 13.2 *Classification criteria for the definition of climatic years*

Classification letter	Criteria
C	Warmest month over 10° C (50° F); coldest month between 3° C (26.6° F) and 18° C (64.4° F).
D	Warmest month over 10° C (50° F); coldest month below 3° C (26.6° F).
ET	Warmest month below 10° C (50° F), but over 0° C (32° F).
a	Warmest month over 22° C (71.6° F).
b	Warmest month below 22° C (71.6° F).
c	Less than four months over 10° C (50° F).
f	No distinct dry season; winter three months (December–February) less than three times as much rain as summer three months (June–August), summer three months less than ten times as much rain as winter three months.
s	Dry summer; winter three months (December–February) at least three times as much rain as summer three months (June–August), and driest summer month receives less than 3 cm (1.2 in) of rain.

Criteria based on G. Trewartha, *An introduction to weather climate* (McGraw-Hill, 1968).

zones is apparent and reflects their barrier effect on the penetration of air masses.

Maps such as Fig. 111 present a more realistic picture of climatic conditions and indicate the difficulty of drawing climatic boundaries.

110 Climatic regions (after G. T. Trewartha)

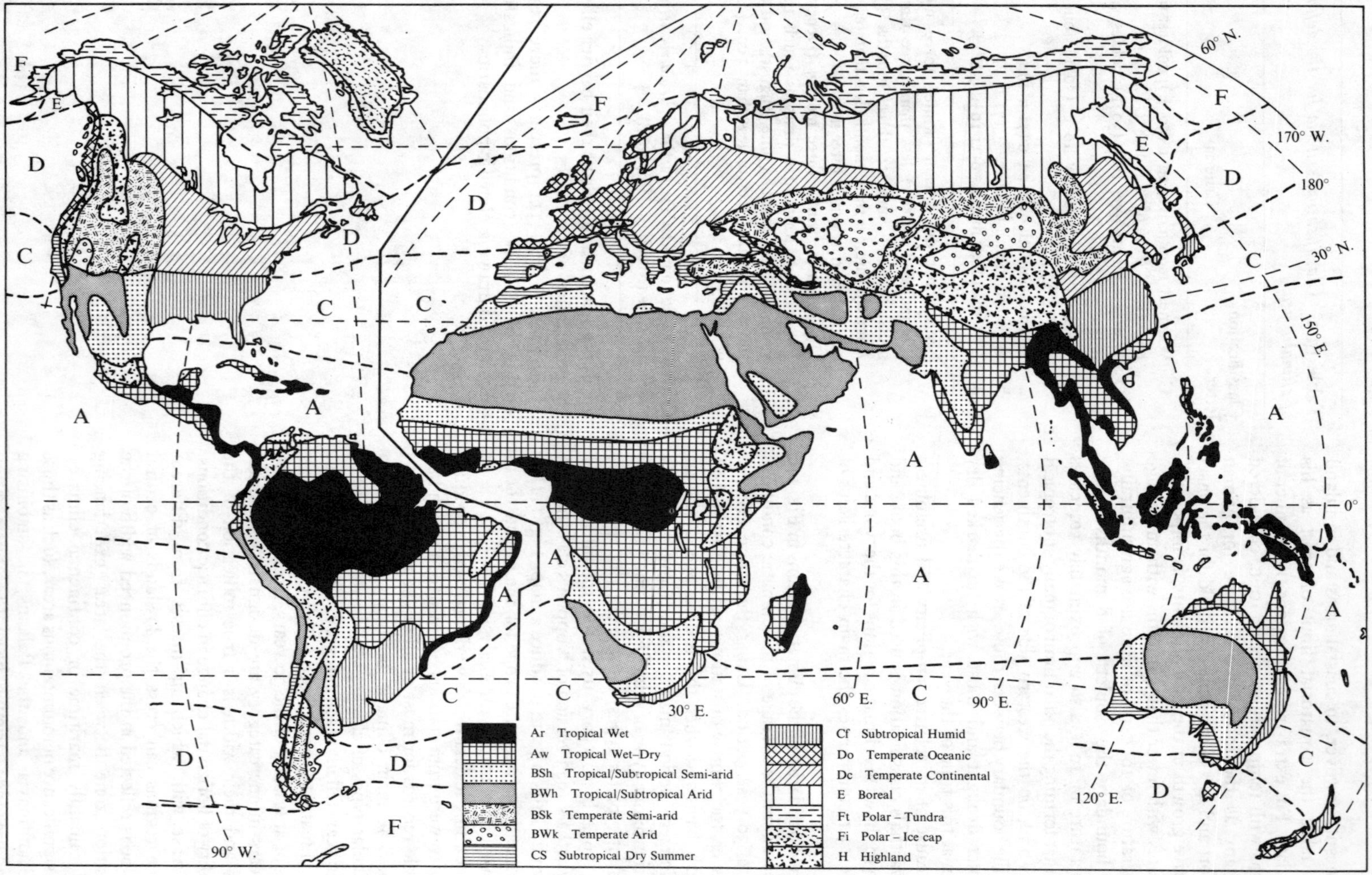
60° N.
170° W.
180°
30° N.
150° E.
0°
90° W.
30° E.
60° E.
90° E.
120° E.
A
C
D
E
F
Ar Tropical Wet
Aw Tropical Wet–Dry
BSh Tropical/Subtropical Semi-arid
BWh Tropical/Subtropical Arid
BSk Temperate Semi-arid
BWk Temperate Arid
CS Subtropical Dry Summer
Cf Subtropical Humid
Do Temperate Oceanic
Dc Temperate Continental
E Boreal
Ft Polar – Tundra
Fi Polar – Ice cap
H Highland

Symbol	Type and main characteristics	Symbol	Sub-types and characteristics	Pressure, wind-belts and main air masses	
				Summer	*Winter*
A	*Tropical humid* No frost. All months over 18°C in coastal areas.	Ar	*Tropical wet* 10–12 months wet	Inter-tropical convergence; doldrums; equatorial westerlies. General wind convergence characteristic. E and mT air masses.	
		Aw	*Tropical wet–dry* Summer rains. Winter – more than 2 dry months.	as Ar	Sub-tropical anticyclones and dry trade winds. cT air masses.
C	*Subtropical* 8–12 months over 10°C; coolest month below 18°C; occasional temperatures below 0°C in inland areas.	Cs	*Subtropical dry summer* Winter half of year has 3 or more times as much rain as summer half. Driest month less than 30 mm. Annual rainfall less than 900 mm.	Eastern sides of subtropical anticyclones – air subsiding and stable. mT and cT mainly.	Mid-latitude westerlies; frontal depressions; anticyclonic spells. mT/mP/cP/cT
		Cf	*Subtropical humid* Rain in all seasons – summer maximum (generally) of convectional character or from tropical cyclones.	Western side of subtropical anticyclones in which air often unstable. Mainly mT.	Mid-latitude westerlies and frontal depressions. mT/cP/mP
D	*Temperate* 4–7 months over 10°C	Do	*Temperate oceanic* Coolest month above 0–2°C. Rain in all seasons with autumn–winter maximum.	Westerlies and frontal depressions dominant. Occasional prolonged anticyclone spells. mP/mT	mP/mT/A/cP
		Dc	*Temperate continental* Coolest month below 0–2°C. Rain in all seasons. Summer maximum of convectional character. Winter snow cover.	Westerlies and frontal depressions. mP/mT	Cold anticyclones. A/cP
E	*Boreal* 1–3 months over 10°C; winters severe; low precipitation – summer maximum.			Westerlies and frontal depressions (mainly occluded). mP/A/cP	Cold anticyclones and polar winds. A/cP
F	*Polar* All months below 10°C.	Ft	*Tundra* Warmest month over 0°C.	Polar winds. Numerous depressions and anticyclones. cP/A	
		Fi	*Ice-cap* Warmest month below 0°C.	Shallow cold anticyclones feed katabatic winds to margins of Antarctica and Greenland. Occasional occluded depressions. A	
B	*Dry* Evaporation greater than precipitation.	BS	*Semi-arid (steppe)* Evaporation greater than precipitation.		
		1. BSh	*Tropical or subtropical* 8 or more months over 10°C. Short, moist season – winter on poleward margins; summer on equatorial margins.	Subtropical high pressure and dry trade winds. Mainly cT.	
		2. BSk	*Temperate* Less than 8 months over 10°C. Low rainfall in summer.	cT	Cold anticyclones. cP
		BW	*Arid (desert)* Evaporation more than twice precipitation.		
		1. BWh	*Tropical or subtropical* 8 or more months over 10°C.	Subtropical high pressure and dry trade winds. cT	
		2. BWk	*Temperate* Less than 8 months over 10°C.	cT	Cold anticyclones. cP
H	*Highland* Altitude affects radiation, temperatures and precipitation.				

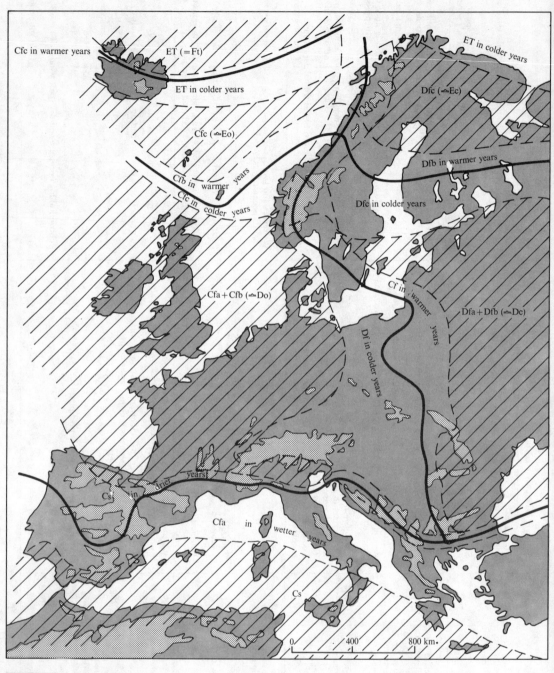

111 Generalised climatic regions of Europe (1901–30), based on climatic year analysis

Cfc in warmer years

ET (= Ft)

ET in colder years

ET in colder years

Dfc (≙Ec)

Cfc (≙Eo)

Cfb in warmer years

Dfb in warmer years

Cfc in colder years

Dfc in colder years

Cfa + Cfb (≙Do)

Cf in warmer years

Dfa + Dfb (≙Dc)

Df in colder years

Cs in drier years

Cfa in wetter years

Cs

0   400   800 km.

 Core areas with 25 years out of 30 with one climatic type

 Land over 950 metres

Transition areas and main climatic boundaries

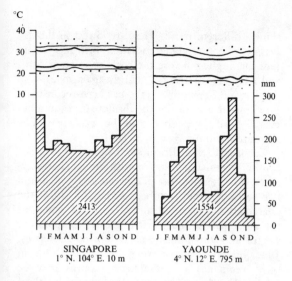

°C

SINGAPORE
1° N. 104° E. 10 m

YAOUNDE
4° N. 12° E. 795 m

The climates of the humid tropics (A)

A wide area on either side of the equator is included in the A climatic zone. Vast areas of the sea; a huge area in the widest part of South America, most of Central America, a broad belt in the narrower waist of Africa and much of India, south-east Asia, the East Indies and the northern littoral of Australia carry the A label with a lower mean monthly temperature limit of 18° c in coastal areas or no frost ever recorded inland. In the centre of the zone all months are wet, but towards the edges there is an increasingly long dry low-sun or winter season. The boundary between Ar and Aw is drawn where there are more than 2 dry months.

An insular station like Singapore (Fig. 112) shows the evenness of the temperature conditions from month to month: no variation more than 1° c on either side of 30° c daily maximum or the 24° c daily minimum. The absolute maximum for any day in a 40-year period is 36° c and the minimum 18° c. Every month has an average of over 150 mm of rain, with the highest month 250 mm.

A station far inland in Amazonia like Uapes on the Rio Negro (not graphed), over 1000 km from the coast and about 2000 km from the mouth of the Amazon, but only 90 m in altitude, is almost as equable. The average daily maximum only varies between 30° and 32° c and the average daily minimum is always within 1° c of 22° c. Being inland, the daily range is 8° c as against 4° c at Singapore. The absolute range is also larger: from 38° c to a low 16° c recorded in a cold snap in July. As seen in chapter 6, cold waves do reach Amazonia from the south in the southern winter. The stronger southern hemisphere circulation has the necessary

push to reach the equator. The average position of the inter-tropical convergence is north of the equator, whereas on a uniform earth it would be positioned on it.

No month at Uapes has less than 125 mm of rain. The wettest month is May with 300 mm, which is about one-ninth of the annual total of 2677 mm.

Yaounde, an inland station in the Cameroons, is an example from Africa. Temperatures are almost as constant as at Uapes and Singapore but the elevation of 795 m reduces all temperatures by 3° c. The absolute maximum is 36° c and the minimum 15° c. The rainfall at this station shows a double maximum of 196 mm in May and 296 mm in October. The maximum falls occur soon after the time of the passage of the overhead sun. In theory this is the pattern that should obtain near the equator while further away from the equator only a single maximum should be found. Actually the occurrence of a double maximum is somewhat capricious. Inspection of rainfall figures suggests it occurs more often to the north of, rather than on, or south of the equator, e.g. Accra 6° N., Akusu 6° N., Ibadan 7° N., Enugu 6° N. show a double maximum, as do Georgetown 7° N. and Paramaribo 6° N. in the Guianas and Colombo 7° N. in Ceylon. This displacement may be due to the asymmetrical position of the inter-tropical trough already noted. The main rainfall at any observation point can be expected when the convergence is in the vicinity of the station. As the convergence moves in consonance, though not in absolute accord, with the overhead sun, the rainy season occurs with high sun in the summer hemisphere and a dry season is experienced at low sun, i.e. in the winter hemisphere.

Kano (Fig. 113) is an example of a station with an Aw climate. Here there are five months with 2 mm or less and of these December is completely rainless. At this season Kano is under the influence of cT air masses which make up the trade winds from the subtropical anticyclone. In Africa these winds are known as harmattan. For a few days in December or January harmattan may penetrate to the Gulf of Guinea coast. From June to September Kano is greatly influenced by equatorial air and the temperature graph for those months is not too dissimilar from that of Yaounde. In the winter months the form of the graph is much more consonant with those of Saharan stations like Agades or even Fort Flatters.

At Kano the average daily maximum is lowest in the rainiest month (August 29° c) and highest before the rains (April 38° c). But average daily minima

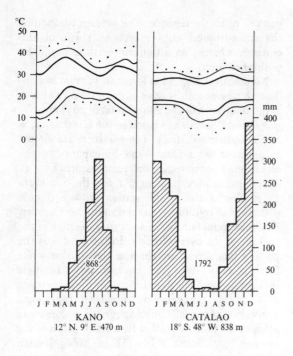

KANO
12° N. 9° E. 470 m

CATALAO
18° S. 48° W. 838 m

show a different trend. They are lowest in the dry season when outgoing radiation brings 13°C in January, highest before the heaviest rains in April and May (24°C) and fall slightly again to 21°C in August and September. 37°C has been recorded in every month and 46°C is the absolute maximum but every April over 40°C can be expected. The absolute minimum is 6°C in January.

Catalao 18° S. in Brazil at a height of 830 m is a southern hemisphere station with similar characteristics. The temperature graph can be divided into two sections: from October to March it shows the characteristics of an equatorial air mass, while for the other half of the year the greater temperature ranges show the influence of trade wind air masses. June, July and August average 10 mm each but in some years any of these three months may be completely dry. Over half the annual total of 1735 mm falls in December, January and February. The average daily maximum varies only from 26°C to 29°C and the minimum from 13°C to 18°C. The absolute minimum is 2°C in June so this station is near the boundary between Aw and Cf.

A variation of Aw is the monsoon type of India which has an even greater proportion of summer rain. Bombay (Fig. 114) has five months (December to April) with 2 mm or less and 95% of its total of 1810 mm is concentrated in the months from June

114 Examples of Aw (monsoon) climate

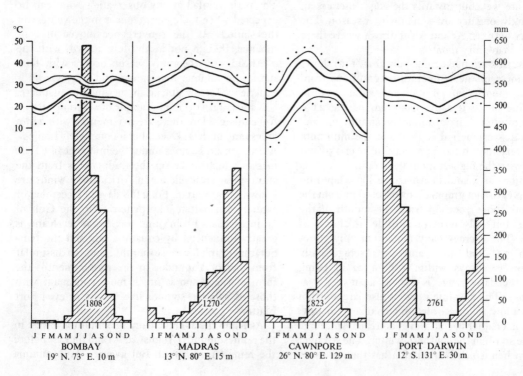

BOMBAY
19° N. 73° E. 10 m

MADRAS
13° N. 80° E. 15 m

CAWNPORE
26° N. 80° E. 129 m

PORT DARWIN
12° S. 131° E. 30 m

to September. Average maximum temperatures vary only 5° C, being highest (33° C) in May before the rains, and reach 32° C after the rains in October and November. January is only 1° C lower than July's 29° C. Minimum temperatures vary slightly more: from 27° C to 19° C. Absolute minimum is 12° C in January, caused by an influx of cold air from the north behind a cold front. March has recorded the maximum of 38° C.

Madras, further south and on the east coast, receives its heaviest monthly rainfall in October and November when over half its total of 1432 mm falls from the retreating monsoon. Maximum temperatures are slightly higher than at Bombay but minimum temperatures are very similar. The absolute range is from 45° C to 14° C.

Cawnpore is an inland station in the Ganges valley. It has seven dry months from November to May when the average fall is only 65 mm, received from the weak western depressions. July and August total 510 mm out of the total of 828 mm. Temperature range is greater. For eight months the maximum is 33° or over with a May average of 41° C. In December and January the average minimum is 8° C and an absolute minimum of 1° C has been recorded and ground frost has occurred. This reflects the incursions of polar air in the rear of the more intense western depressions.

Port Darwin in north Australia has a monsoon climate. Slightly nearer the equator than Madras, the range of maximum temperatures is from 31° C to 33° C and the minimum 21° C to 27° C. 37° C has been reached in every month but 41° C is the upper recorded limit. There are five dry months with June, July and August recording 2 mm or less. 1224 mm of the total of 1500 mm fall from December to March.

Aw climates have a high unreliability of rainfall and suffer from flood and drought.

Arid climates (B)

The tropical, subtropical, temperate, boreal and polar climatic zones are defined firstly by temperature characteristics and subdivided by amount and season of precipitation. Dry climates are defined as areas where water losses exceed the gains from precipitation and are subdivided by their temperature characteristics. They cut across the zonal arrangement of A C D E F.

Climatologists have tried to formulate rules which define the boundaries of arid climates. One of the simplest, using only temperature and rainfall, is

due to Köppen, the great German geographer. The annual precipitation in millimetres (P) and the annual mean temperature in degrees Celsius (T) are required. In areas of predominantly winter rain, if P is divided by T and the result is 10 or less the climate can be considered arid. If the result is between 10 and 20 it is semi-arid: if above 20 it is outside the arid boundary.

Here are some station figures to show the method of application:

	P	T	P÷T	Category
Athens	395	17	22.6	Sub-humid
Alexandria	206	21	9.8	Arid
Morocco	240	19	12.6	Semi-arid
Santiago (Chile)	360	13	27.7	Sub-humid
Adelaide	540	17	31.8	Sub-humid
San Diego	250	16	15.6	Semi-arid
Cairo	30	21	1.4	Arid
Baghdad	180	26	6.9	Arid

In areas of summer rain more rainfall is necessary because temperatures are higher in the rainfall season. P has now to be divided by T + 14.

	P	T	P÷ (T+14)	Category
Alice Springs	251	21	7.2	Arid
Karachi	190	26	.8	Arid
Kayes	744	30	17.0	Semi-arid
Bulawayo	640	19	19.4	Semi-arid
Iguata (Brazil)	787	28	18.7	Semi-arid
Ulan Bator	208	−3	19.0	Semi-arid

Where there is no clearly defined season of maximum rain the divisor used is T + 7.

	P	T	P÷ (T+7)	Category
Kamloops	265	8	17.6	Semi-arid
Cipolleti (Argentine)	162	15	7.4	Arid

Calculations of this kind serve to delimit areas of true aridity from the semi-arid ones. Areas of extreme aridity are those which have a record of no rainfall at all for twelve consecutive months. Fig. 115 illustrates the climatic variations found within the arid regions experiencing a BWh climate.

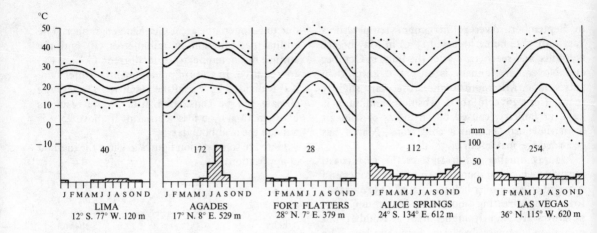

LIMA
12° S. 77° W. 120 m

AGADES
17° N. 8° E. 529 m

FORT FLATTERS
28° N. 7° E. 379 m

ALICE SPRINGS
24° S. 134° E. 612 m

LAS VEGAS
36° N. 115° W. 620 m

Lima in Peru has only 40 mm average annual precipitation and every month has a mean daily maximum of 17° C or above. The lowest daily minimum is 13° C and the lowest daily maximum, also in August, is 19° C. The coastal position and the cold current keep temperatures and range low. 34° C is the highest ever recorded and the absolute low is 9° C.

Agades (17° N.) in Africa shows the contrast at an inland site. Most of the total of 172 mm falls on a few days in July and August and the six months from October to February are completely dry. Average daily maximum temperatures reach 42° C and even 30° C in the coldest month. The daily range is large, often 20° C. 4° C is the absolute minimum.

Fort Flatters (28° N.) has hardly any rainfall – just 28 mm on average. What rain there is falls mainly in the winter months. There is a large daily range of about 17° C and an annual range of 23° C. In July average daily maximum temperatures are 43° C, and 51° C has been recorded. Minimum temperatures of freezing and below are reached every year in December, January and February and a low of −7° C has been recorded. The periodic winter incursion of polar air masses into the interior of North Africa and the Middle East partly accounts for the low temperatures experienced.

Alice Springs (24° S.) in the centre of Australia has a summer rainfall maximum with 191 mm of the 251 mm total falling from October to March inclusive. Daily minima below freezing are recorded regularly in June and July and frost has occurred from May to September. Average daily maximum temperatures vary from 36° C in January to 19° C in July.

All these four stations are arid (BW). Fort Lamy with a mean annual temperature of 28° C and 746 mm

precipitation is semi-arid, while Kano is quite near but just outside the semi-arid boundary zone. On the northern edge of the Sahara, Tripoli (19° C and 385 mm) is just on the Bsh–Cs boundary. The transition from Cs on the Mediterranean coast to arid inland is very rapid.

In the U.S.A. there is a large area of desert and semi-desert in the mountain-girt basin and range states of the Rockies. Las Vegas has a low rainfall total of 112 mm and a 20° C temperature range between 8° C and 28° C. The daily range is not much less than this. Both January and December have average daily maxima of 15° C and minima of −2° C. July and August have very hot days (39° C average maximum) but much more reasonable nights of 20° C. There are five months with some frost and low temperatures due to radiational night-time cooling, particularly when cP air masses invade the area. The air is generally dry. Conditions are arid (Bwh) at Las Vegas (18° C and 112 mm), semi-arid (Bsh) at Phoenix (21° C and 396 mm, predominantly in summer) and also semi-arid but Bsk at Cheyenne (7° C and 377 mm, with a summer maximum). Strict Bwk conditions are hardly realised in the U.S.A. but there are large areas in central Asia, e.g. around Krasnovodsk on the eastern shore of the Caspian. Ashkabad (16° C and 225 mm, with predominantly winter rain) is semi-arid, as is Ulan Bator further east and north with −3° C and 208 mm, falling mainly in summer. Everywhere in these dry climates temperature contrasts between day and night and summer and winter are great, but in central Asia the extremities are increased by the size of the land mass, the elevation and the dominance of cP air masses. Ulan Bator at 1345 m has a range from 16° C to −26° C and ground frost has occurred in every month.

Subtropical climates

These are transitional between those of lower and higher latitudes, being mainly under the influence of the mid-latitude westerlies in winter and the subtropical anticyclones in summer.

The subtropical dry-summer or Mediterranean (Cs) climates

Located on the west coastal margins between latitudes 30° and 40° (30° to 45° in Europe/north-west Africa), their most distinctive characteristic is the summer drought/winter rainfall regime. Each of the individual areas, however, possesses some unique climatic characteristics arising from its location and overall physical geography.

The southern hemisphere

Figure 116 illustrates the regimes in the southern hemisphere. The low summer rainfall of Valparaiso, representative of the coastal area of central Chile, results from the anchoring of the eastern end of the semi-permanent subtropical anticyclone by the Andes and the presence of a pronounced inversion at 1000–1500 m. The low summer temperatures are the result of the upwelling of the cold waters of the Peru current with a surface temperature of 15°C (10°C in winter). In winter the subtropical anticyclone migrates northwards and spells of rain occur with the passage of cold fronts of Pacific depressions. On average, Valparaiso has only 31 raindays a year, of which 24 occur in the four mid-winter months.

In the South Atlantic and Indian Oceans the subtropical anticyclones migrate eastwards (see pp. 61–3) and the cold frontal troughs between them give occasional summer rain in the Cs climates of South Africa and Australia. The appreciably higher winter rainfall in south-west Australia (Perth graph Fig. 116) than in south-west Africa or central Chile is due to the ocean to the south and west of it being an area where cyclogenesis occurs. Most of the rainfall is in the form of showers which occur in the mP air behind the cold fronts.

The large mean diurnal temperature ranges in all localities in the southern hemisphere reflect the preponderance of clear sky conditions; the high monthly and absolute maxima of summer in South Africa and Australia the periodic incursions of cT air from the nearby deserts. The greatest diurnal temperature ranges occur during such incursions, often being some 20°C to 25°C in winter and more in summer.

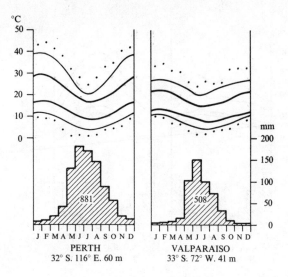

PERTH
32° S. 116° E. 60 m

VALPARAISO
33° S. 72° W. 41 m

North America

In California (Fig. 117) the summer drought is as pronounced as in central Chile, due to the anchoring of the Hawaiian anticyclone by the western Cordillera and again the number of raindays is small, increasing northwards from 37 at Los Angeles to 67 at San Francisco and 120 at Eureka.

For the purpose of giving a better overall picture of climatic conditions within a particular area than can be obtained from the climatic statistics of individual stations the United States Weather Bureau has calculated the averages for each physiographic subdivision within each state. Most of the subdivisions are drainage areas for which the data are valuable in respect of flood control and irrigation.

Figure 118 shows the average rainfall for the San Joaquin drainage area of central California for each of the water years (October to September) from 1931–2 to 1959–60. The rainfall for the wettest month in each winter is also shown. The driest winter (1958–9) received 55% of the normal; the wettest (1937–8) 157%. The fact that 10 of the first 14 winters were wetter than average but only 4 of the next 15 indicates a possible periodicity which must be taken into account in planning for water storage to meet the long-term needs of an area such as this.

The concentration of the rainfall in a relatively few days each year has already been mentioned. A large proportion of the annual rainfall can also be concentrated in a period of days of short duration. This is also indicated in Fig. 118, which shows that the contribution of a particular month to the winter (and annual) total varied from *c.* 23% (April 1934–5) to 52% (January 1952–3).

The range of temperatures which are recorded

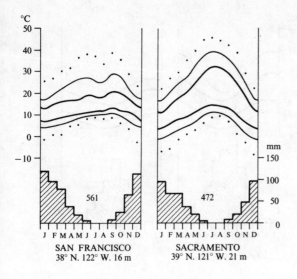

SAN FRANCISCO
38° N. 122° W. 16 m

SACRAMENTO
39° N. 121° W. 21 m

118 Variations in rainfall for San Joaquin drainage area, 1931–60

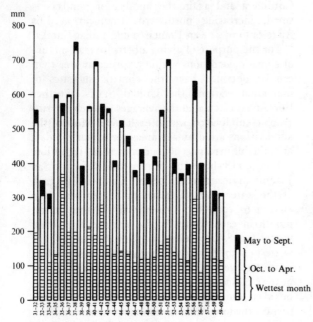

(Fig. 117) show the marked contrast between coastal localities affected by the cold Californian current (San Francisco), and the more continental regime of the central valley (Sacramento). In the coastal area the low mean daily and monthly maxima, the levelling out of the temperature graphs between May and August (when the main upwelling of cold water occurs with the persistent northerly winds) and occurrence of the peak monthly temperature in August/September all reflect the chilling effect of the current. The low surface temperatures accentuate the semi-persistent anticyclonic inversion of the summer ridge of high pressure below which noxious fumes are trapped to form a major health hazard. The current is also responsible for the coastal fogs which are particularly frequent along the northern coast as well as for the negative temperature anomalies.

The Mediterranean Sea area
The Cs climate of this area has many facets due to the complex configuration of the coastline of the warm sea extending some 3000 km into the heart of a land mass with great variations in altitude and aspect and to the fact that the great majority of depressions result from cyclogenesis within the region and leave by a number of preferred tracks (Fig. 64).

Figure 119 shows the variation in the season of maximum rainfall. Only in the southern and eastern parts of the Mediterranean (Algiers and Haifa graphs) does a relatively simple regime of winter rain/summer drought occur. The southern-most areas also feel the full effects of incursions of cT air and rarely experience frost. Conversely the northern coastlands, affected by periodic incursions of cP and cA air in winter, frequently experience temperatures below freezing point (Split, Marseille and Rome graphs). Whilst the western part of the Mediterranean is dominated by a ridge from the Azores anticyclone in summer, the east is under the influence of the monsoonal low pressure system of the middle East and of the northerly etesian winds which account for the drop in the absolute and mean monthly maximum temperatures recorded at Haifa in midsummer.

The double rainfall maximum recorded by some northern stations, e.g. Marseille, indicates a transition to the continental climate of central Europe. The more pronounced October/November maximum is due to the warmth of the sea which results in the air having a high absolute humidity at a time when cyclogenesis is marked.

119 Examples of Cs climate (Mediterranean Sea area)
120 (*below*) Variability of rainfall, Cagliari, Sardinia,
1951–60

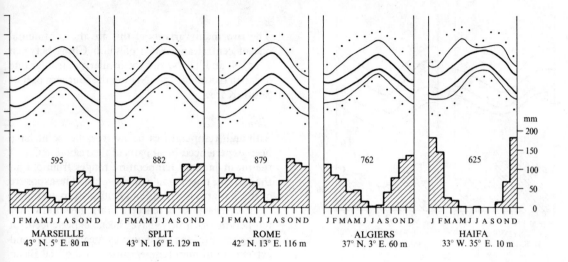

MARSEILLE
43° N. 5° E. 80 m

SPLIT
43° N. 16° E. 129 m

ROME
42° N. 13° E. 116 m

ALGIERS
37° N. 3° E. 60 m

HAIFA
33° W. 35° E. 10 m

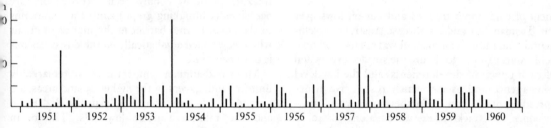

The marked intensity of the rainfall is reflected in the small number of raindays in the wetter areas, e.g. Rome 73, Split 82. The heavy winter falls are associated with marked instability along the cold fronts and in the cold air behind them due to the sea temperatures being some 2° C higher than those of the invading cold air masses. The Mediterranean air in the warm sectors of depressions also tends to be unstable. Particular heavy falls occur when surface lee depressions, forming over the Gulf of Genoa and to the south of the Atlas mountains, deepen rapidly when overrun by an upper cold cut-off low. Such an occurrence caused the torrential rains and floods in central and south Tunisia and north-east Algeria at the end of September 1969 when Biskra recorded 246 mm during the period 20–29th, of which 210 mm fell on the 27th and 28th. Biskra's mean September rainfall is 17 mm and the mean annual 148 mm.

Above average cyclonic activity gives abnormally wet months or years whilst below average gives dry ones. The variability of the rainfall is shown in the graph for Caglihari (Fig. 120), where the extreme annual totals for the period 1951–6 were 275 and 738 mm. During the same period Marseille recorded extremes of 388 and 803 mm and Rome ones of 413 and 1027 mm.

Subtropical humid (Cf) east coast climates

In contrast to the Mediterranean (Cs) climates most areas with a Cf climate have a summer maximum of rainfall associated with the influx of humid mT air from the unstable western ends of the oceanic subtropical anticyclones. Variations between one region and another occur, however, because of the effects of local geographical conditions on the atmospheric circulation.

In the southern hemisphere these climates are of limited extent in south-east Africa and south-east Australia because of the topographic barriers of the Drakensberg mountains and Great Dividing range respectively. In Australia the dominance of maritime air masses is reflected in the generally lower mean daily and monthly temperature ranges experienced at Sydney (Fig. 121) compared with those for Perth with its Cs climate. The absence of a dry season is associated with the all-year-round develop-

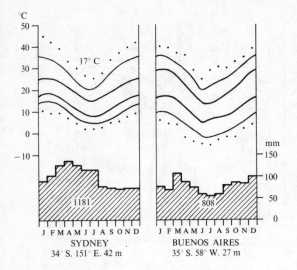

SYDNEY
34° S. 151° E. 42 m

BUENOS AIRES
35° S. 58° W. 27 m

ment of cold upper troughs and cut-off lows over the Tasman Sea and south-east Australia. Cyclogenesis and the development of waves on meridional cold fronts east of 130° E. are the surface expression of such upper air developments, and the low level convergence and ascent which result give widespread general rains. Furthermore when anticyclones are tracking eastwards along c. 40° S. onshore humid easterly or south-easterly winds on their northern side are subjected to orographic uplift by the Great Dividing range, and, if the air is latently unstable, outbreaks of rain occur.

The seasonal distribution of rainfall in the south shows a maximum in mid-winter when the westerlies and their associated depressions are on their most northerly tracks. Further north (Sydney graph, Fig. 121) the maximum normally occurs in autumn when tropical cyclones, which give particularly heavy falls, affect the area.

Unlike the coastal regions of south-east Australia and south-east Africa the Cf climatic region of South America experiences winter frosts (Buenos Aires graph, Fig. 121). These occur in the outbursts of antarctic air, called the *frigorem*, across Patagonia. The high midsummer maxima occur in outbursts of cT air from the Argentinian desert. The Pampas, unlike south-east Australia, has only a few rainy days (Buenos Aires 63, Sydney 152) despite the presence of a frontogenetic zone running southeastwards from the river Plate. This is due to the semi-permanent upper ridge with its axis east of the Andes (see p. 61) which inhibits convection. When rainfall does occur with the passage of a cold front or due to instability it is of marked intensity.

East Asia
The two main variants of this are the continental one of central and much of south China between c. 25° N. and 34° N. and the insular/peninsular one of Japan (south of c. 38° N.) and the southern part of South Korea.

China
Although temperatures below freezing point have been experienced in all parts of the region in China the mean January temperature ranges from 0° C in the extreme north to 14° C in the south. The exposure of the Yangtze lowlands and delta area to outbursts of cP air in winter is reflected in the low midwinter temperatures which have been experienced in Shanghai (Fig. 122) as shown by mean monthly and extreme minimum temperatures. In the Red Basin of Szechwan on the other hand temperatures below freezing point have only been recorded on rare occasions (Chungking graph) since the mountains to the north form a barrier to the ingress of cP air which is warmed adiabatically on the rare occasions it does penetrate.

Mean midsummer temperatures are remarkably uniform and, except in the higher upland areas, are over 28° C but only exceed 30° C in the Red Basin and the middle Yangtze lowlands. The mean monthly and extreme maximum temperatures reveal the periodic influx of very warm tropical air masses which in midsummer give temperatures which occasionally exceed 40° C.

The mean annual rainfall ranges from 750 mm in the extreme north, which averages 75 raindays a year, to over 2000 mm in the uplands of north Fukien and Kwangsi, where the rain days exceed 150. The Red Basin is anomalous in that parts of it receive over 1500 mm – up to 500 mm more than the surrounding upland – a reflection of pronounced and frequent convection during the summer.

As in tropical south China, most parts of the subtropical region have a simple seasonal rainfall regime with a minimum in midwinter followed by a marked increase in March, April and May, as the moisture content and instability of the air and cyclonic activity increase. Everywhere the maximum occurs in summer. Over the south China uplands the summer rainfall is partly cyclonic and partly associated with convergence and pressure surges within the mT air stream from the Bay of Bengal which has a high absolute humidity. In the Yangtze valley (Shanghai graph Fig. 122) and parts of the south-east the June maximum is given by the

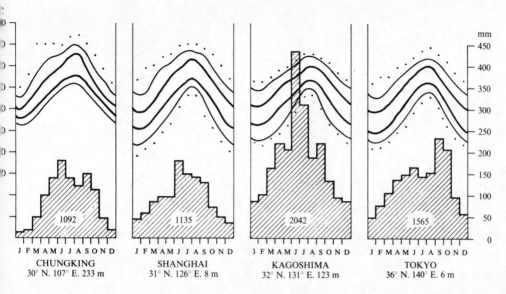

mm

CHUNGKING
30° N. 107° E. 233 m
1092

SHANGHAI
31° N. 126° E. 8 m
1135

KAGOSHIMA
32° N. 131° E. 123 m
2042

TOKYO
36° N. 140° E. 6 m
1565

Mai-u frontal rains (see p. 70); in the eastern coastal region the maximum is in July or August when typhoons make a marked contribution, e.g. Nanking receives 18% of its annual total in August. Typhoons are estimated to account for 30–35% of the annual rainfall on the south China coast (Ar climate), 10–20% on the east coast and 5% in inland areas.

The rainfall starts to decrease markedly in late September and October as outbursts of cP air increasingly take place. Some areas, e.g. the Red Basin (Chungking graph Fig. 122), experience a secondary maximum (Shurin rains) as the main frontal systems move south again.

Japan
Despite the insularity of Japan mean winter temperatures are lower along the Pacific coast line than those experienced in comparable latitudes on the coast of the United States (e.g. January: Osaka 4° C, Wilmington 8° C). The winter temperatures along the south coast in fact only approximate to those in south-west England, despite the difference in latitude and despite sea temperatures of 10–12° C on the continental side of the country and 13–14° C on the Pacific side. The low winter temperatures pinpoint the regularity of the winter monsoon which gives frequent incursions of cP air, although it is periodically interrupted by the passage of depressions. Fig. 122 shows that the area has a negative winter temperature anomaly in common with the rest of east Asia. Since summer temperatures are high, due to the low latitude, the annual temperature ranges, although somewhat

smaller than on the mainland, are of continental proportions.

The insularity of the climate is mainly seen in the absence of a dry season (Tokyo and Kagoshima graphs). The maximum of early summer corresponds to the period of the Bai-u frontal rains and the onset of the typhoon season, the second maximum of early autumn to the Shurin rains, but again includes rainfall from typhoons.

South-east U.S.A.
Occupying an area some 2000 km from west to east and 1000 km from south to north, Cf gives way to the semi-arid climate along *c*. 100° W. and to the temperate continental along *c*. 40° N.

Although mean January temperatures are generally above freezing point and reach 18° C in the south, marked fluctuations of temperature occur with the passage of anticyclones and depressions, as the graphs in Fig. 123 show. Particularly noticeable are the cold spells due to the ease with which cP air penetrates southwards to give average annual minimum temperatures of −23° C in the north and 0° C in the south. Average summer temperatures are much more uniform, ranging from 29° C in south Texas to 24° C in the north, whilst average annual maximum temperatures range from 32° C on the Atlantic coast to 40° C in the interior.

Mean annual rainfall ranges from 500 mm in the west to 1250 mm on parts of the Atlantic coast, 1625 mm on parts of the eastern Gulf coast and 2000 mm in the south Appalachians. Appreciable variations occur from year to year, however, especially in the part bordering the semi-arid region,

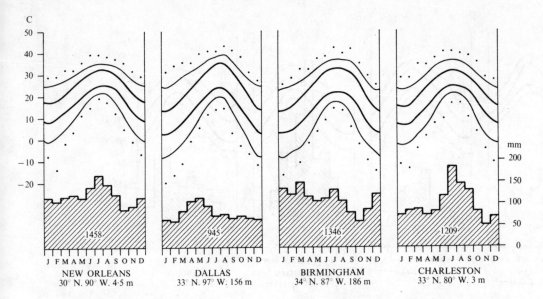

NEW ORLEANS
30° N. 90° W. 4·5 m

DALLAS
33° N. 97° W. 156 m

BIRMINGHAM
34° N. 87° W. 186 m

CHARLESTON
33° N. 80° W. 3 m

as the figures for the period 1931–60 for three drainage areas indicate.

	Low rolling plains of Texas (western boundary)	Mississippi delta	Southern South Carolina
	mm	mm	mm
Average	584	1558	1182
Wettest year	1125	1915	1675
Driest year	331	1221	734

There is no pronounced dry season but variations do occur in the seasonal distribution. On the Atlantic and Gulf coastal plains east of 95° W. (Charleston and New Orleans graphs) there is abundant precipitation throughout the year from frontal depressions but the noticeable summer maximum is due to the summer thunderstorms and hurricanes. Over and near the Florida peninsula the maximum extends into autumn when hurricanes are most frequent.

The subtropical interior east of 95° W. (Birmingham graph) has a rainfall maximum in the cool season when frontal depressions are most frequent. A secondary maximum of a convectional character occurs in July. West of the Mississippi the maximum occurs in late spring (Dallas graph) when Pacific air masses seldom extend beyond the Rockies and the interaction of mT and cP air masses gives disturbed weather.

Temperate oceanic (Do) climates

Oceanic climates of middle latitudes can be broadly defined as those areas with a mean annual temperature range of less than 20° C and experiencing maritime air masses for at least 75% of the year. Because of the prevalence of the westerly air flow they are limited to the western margins of the continents, especially in North and South America where the meridionally orientated mountain ranges favour the development of upper ridges (pp. 9, 59) on the western (coastal) flanks of which maritime air masses and mid-latitude depressions are transported polewards. Only in western Europe where the north European plain allows easy penetration of North Atlantic air masses and depressions is the oceanic climate extensive.

Figure 124 illustrates the climatic conditions in each of the southern hemisphere areas experiencing a temperate oceanic climate. The persistence of a strong westerly air flow south of 40° S. is reflected in the small temperature ranges and high monthly rainfall totals experienced at west coast stations (Hokitika). As the boundary with the Cs climates is approached a higher proportion of the precipitation is concentrated in the winter months. The pronounced influence of the Humboldt cold current on temperatures along the Chilean coast is reflected in the fact that the *midsummer* temperatures at Los Evangelistas in southern Chile approximate to the *midwinter* temperatures at Valentia in south-west Ireland which is in a comparable latitude.

The graphs for Hokitika and Christchurch clearly illustrate the intensification of rainfall on the westward side of mountain ranges and the rapid

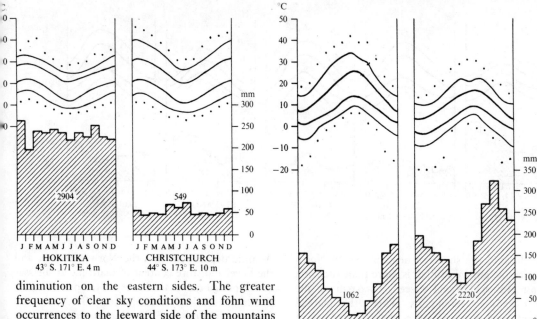

HOKITIKA
43° S. 171° E. 4 m

CHRISTCHURCH
44° S. 173° E. 10 m

PORTLAND
46° N. 123° W. 47 m

SITKA
57° N. 135° W. 5 m

diminution on the eastern sides. The greater frequency of clear sky conditions and föhn wind occurrences to the leeward side of the mountains accounts for the higher maximum temperatures experienced especially in summer at Christchurch, where the secondary rainfall maximum in December and January is due to the development of convection over the Canterbury plains.

The difference between the mean monthly temperature of a particular place and the average for the latitude is called an *anomaly*. In the southern hemisphere, where oceanic influences are dominant, the anomalies are generally small, the most persistent being the negative ones on the west coastal areas of Africa ($-4°$ c) and South America ($-4°$ c) due to the Benguela and Peru cold currents.

The much greater temperature anomalies in the northern hemisphere, especially in winter, reflect both the greater extent of the land masses, which in middle and high latitudes undergo pronounced radiational cooling in winter and heating by insolation in summer, and the poleward transportation of warmth by the warm currents and mT air masses of the North Pacific and North Atlantic, especially in winter. The deep, dynamically induced low pressure areas of the Aleutians, to the lee of the Kamchatka peninsula, and of Iceland, to the lee of Greenland, give rise to a strong flow of air which transports warmth and moisture towards the western sides of North America and, especially, Europe. The fact that in Europe a positive winter anomaly is experienced to *c*. 40° E. in the vicinity of the Black Sea and to *c*. 80° E. on the arctic coast of Russia and as high as $+20°$ c in the Norwegian Sea emphasises the relative ease with which maritime

air masses penetrate the continent. The lower positive anomalies over the north-east Pacific ($+12°$ c) in winter reflect the fact that the Alaskan current is cooler than the North Atlantic Drift.

The temperature graphs for the northern hemisphere stations in Figs. 125 and 126 show the much greater ranges experienced compared with the Do climates of the southern hemisphere. These are partly caused by the greater amplitude of the Rossby waves which result in pronounced meridional exchanges of air, which give rise to the marked temperature changes experienced in the short alternating warm and cold spells of weather.

The low mean monthly and absolute minimum temperatures recorded in winter occur in outbursts of arctic and continental polar air and are particularly pronounced when night-time radiational cooling of surface air takes place under the clear-sky and slack pressure-gradient conditions associated with cold anticyclones. Conversely the high maxima recorded in summer are associated with intense day-time heating under warm anticyclonic conditions and the occasional influx of cT air from the Sahara into western Europe and from intermontane deserts of south-west U.S.A. to the Pacific coastlands.

The rainfall regimes of the Do climates of the northern hemisphere are somewhat more complex

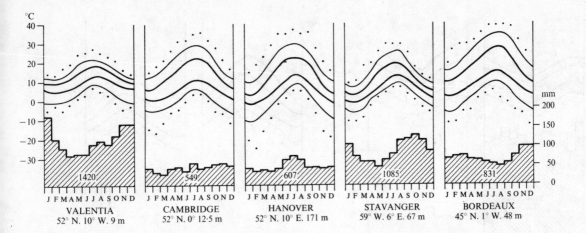

than those of the southern hemisphere. In North America the annual rainfall is more than 1000 mm almost everywhere and exceeds 6000 mm on the higher exposed mountains. There the marked concentration of precipitation in the winter half of the year emphasises the much greater frequency and intensity of depressions compared with summer when frequent ridges from the Hawaiian high dominate the weather. South of 50° N. the peak precipitation occurs in midwinter (Portland graph, Fig. 125); to the north in autumn (Sitka graph, Fig. 125). The autumn maximum in the northerly part of the area is due to the slow fall of sea surface temperatures compared with those of the land and the consequent increase in orographic rainfall from the warm, moist onshore winds (cf. Stavanger in north-west Europe). Although the number of days with snow at sea level is small, the fact that the midwinter snowline is only at 200–300 m in British Columbia results in heavy winter snowfalls. Thus in the Kildare Pass (1600 m) south-east of Kitimat the average winter snowfall is 20.55 m.

In western Europe, where the Do climate extends further into the continent, the rainfall regimes are less uniform. Along the Atlantic coast a pronounced rainfall maximum occurs during the winter half of the year, the peak varying from October (e.g. at Stavanger) to January (e.g. at Valentia) (Fig. 126). This is a period when westerly and cyclonic weather systems are dominant, except in mid-November when anticyclones frequently develop. The deep and vigorous nature of the depressions and the marked instability of mP air in autumn when the sea temperatures are at their highest accounts for this marked maximum. The spring/early summer minimum of precipitation reflects both the increased frequency of blocking anticyclones which divert

Atlantic depressions into the Norwegian Sea, and the lower moisture content of maritime air masses at this period due to the lower sea temperatures. At the more southerly stations (e.g. Bordeaux) the frequent influence of the Azores ridge of high pressure and the higher latitudinal tracks of depressions are reflected in the noticeable summer minimum.

Many inland stations on the European mainland (e.g. Hanover) and some in eastern Britain (e.g. Cambridge) experience a maximum of precipitation during the summer months following a sharp increase of westerly and north-westerly weather frequency around mid-June. These invasions of maritime air are sometimes called the 'summer monsoon' of Europe. In these air streams convectional instability occurs following day-time heating and is the cause of the summer maximum. There is normally a decrease in precipitation in mid-September when a spell of anticyclonic weather frequently occurs.

High annual rainfall totals are recorded in the mountainous areas of Europe as a result of orographic precipitation, the increased shower activity and the intensification of fronts on the windward slopes. The lowland areas between 45° and 55° N., however, have generally a low total rainfall. There are three reasons for this: (a) it is the zone least crossed by the centres of depressions; (b) the depressions and fronts crossing the area are of only moderate intensity because of the generally small differences in temperature and density between the mP and mT air masses; (c) many of the fronts are occluded by the time they reach western Europe and as a result the frontal rain is usually light in character.

The graphs in Figs. 123, 125 and 126 show the

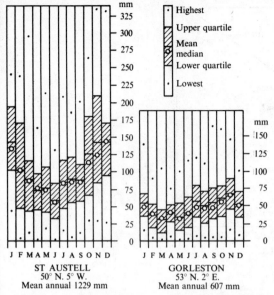

mm
- Highest
Upper quartile
Mean
median
Lower quartile
- Lowest

ST AUSTELL
50° N. 5° W.
Mean annual 1229 mm

GORLESTON
53° N. 2° E.
Mean annual 607 mm

range of temperatures to be expected in any particular month, but do not indicate the range of precipitation to be expected. Fig. 127 illustrates the variability of rainfall for two places in England – St Austell in Cornwall and Gorleston on the east coast of Norfolk for the period 1931–60.

The southern coastal areas of south-west England have relatively dry summers because they are in the part of the country most frequently under the influence of ridges of high pressure from the Azores anticyclone. The upland areas of western Britain have less rainfall in summer than in winter since frontal depressions are fewer and less intense and the mP air in their rear more stable.

The eastern parts of Britain which receive a summer maximum of rainfall are the areas most affected by convectional showers at this season of the year – a reflection of the 'continentality' of the climate.

The graphs show the variability of the rainfall amounts to be expected in any month and are a good indication of the variability of British weather. The interquartile values show the range of rainfall received in 50% of the years. From these graphs one can calculate the variations from the mean. Thus St Austell has a mean rainfall in March of 89 mm. The interquartile values range from 36% (lower quartile) to 130% (upper quartile) of this. On one occasion in this period, however, the March total was 333% of the mean and on another only 16%. Such information is important in many aspects of everyday life and must be taken into consideration in considering the demands of water for domestic and industrial purposes and the need and feasibility of drainage and irrigation schemes.

In essence the Do climate can be summarised as one of marked variability. Within the span of a few days the alternation of air masses can cause a change from mild to cold weather in winter and from hot to cool in summer (or vice versa); two or three wet or very wet days can be followed by a dry spell which may be cloudy or sunny, warm or cold, clear or foggy. The persistence of a particular synoptic situation can result in a particular season being cooler or warmer, wetter or drier than average and over the longer term whole years can be characterised in these terms. Such are the results of lying within latitudes which experience the main meridional interchange of air between the tropical and polar regions.

The temperate continental (Dc) climates

In North America and Scandinavia the transition from maritime to continental climates is rapid, due to the barrier imposed on the free inland flow on the surface layers of maritime air masses from the North Pacific and North Atlantic respectively and to the decrease of temperature with altitude. Thus along the 160 km long Sogne Fiord, Norway, the mean January temperature decreases from 2° C at the mouth to − 1.1° C at the head, and mean annual precipitation from 1500–2000 mm to 500 mm. The climatic regime, however, is maritime along the whole length of the fiord. The mountain watershed lies only c. 25 km east of the head of the fiord and in the Ottadal valley to the east of it mean January temperatures are −7° c to −9° c.

Across the European plain the transition from maritime to continental conditions is much more gradual but continentality across Eurasia, as measured by the increase in mean annual temperature range and percentage of precipitation falling in the summer half of the year (Fig. 128), increases until the Pacific coast is reached.

The graphs in Fig. 129 illustrate the main features of the Dc climate of Europe. The area is dominated by cold continental air in winter – hence the low mean daily maximum and minimum and mean monthly minimum temperatures resulting from the snow cover, dry air and atmospheric stability. Periodic incursions of maritime air and filling frontal depressions from the Atlantic and Mediterranean are responsible for the milder spells indicated by the mean monthly and absolute maximum temperatures and for the not inappreciable winter precipitation.

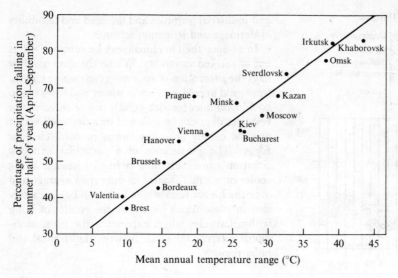

129 Examples of Dc climate (Europe)

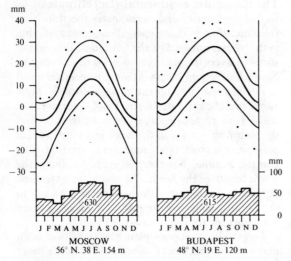

MOSCOW
56° N. 38 E. 154 m

BUDAPEST
48° N. 19 E. 120 m

130 Frequency of weather classes, Moscow (after Federov)

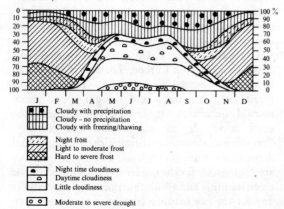

Cloudy with precipitation
Cloudy – no precipitation
Cloudy with freezing/thawing

Night frost
Light to moderate frost
Hard to severe frost

Night time cloudiness
Daytime cloudiness
Little cloudiness

Moderate to severe drought

Features of all continental climates are the rapid transition from winter to summer temperature conditions, and the influx of maritime air is generally low pressure conditions become dominant. These are accompanied by the increased instability which accounts for the summer rainfall maximum. The high mean monthly and absolute maxima of temperature in summer reflect the intensity of insolation. Heatwaves are particularly associated with invasions of cT and mT air.

Figure 130 provides a succinct summary of the weather conditions experienced at Moscow and highlights the contrasts between winter and summer and the shortness of the intermediate seasons.

The Dc climate of mainland east Asia, especially the north, is even more dominated by cP air in winter than that of Europe. This is reflected in the low mean daily maximum and minimum temperatures and low rainfall at Vladivostok (Fig. 131). Even so, a wide range of temperatures can be experienced, the warmer days on the coast occurring when the wind is southerly and skies overcast. The winter precipitation is associated with frontal depressions skirting the coast. The rainfall graph clearly brings out the summer monsoonal influences. During summer the cooler and cloudier days occur when southerly winds are chilled by the cold waters of the Okhotsk current.

The climatic graph for Akita illustrates the climatic conditions experienced on the western side of the offshore islands. Much of the winter precipitation is orographic and the result of the

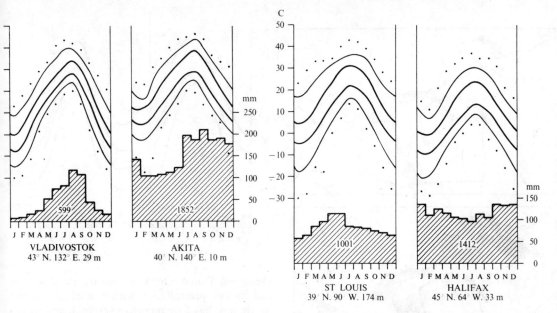

VLADIVOSTOK
43° N. 132° E. 29 m

AKITA
40° N. 140° E. 10 m

ST LOUIS
39° N. 90° W. 174 m

HALIFAX
45° N. 64° W. 33 m

conditional instability developing in the cP air warmed and moistened in its passage over the Sea of Japan, despite the presence of the cold Okhotsk current whose influence is reflected in the low temperatures. The sudden increase in precipitation in June is due to the Bai-u rains, which are followed by the typhoon period and the Shurin rains.

Figures 4 and 132 illustrate the conditions in the Dc climate of North America. The ease with which tropical air masses can penetrate northwards and arctic air masses southwards is particularly reflected in the ranges of temperature recorded at Edmonton and St Louis.

The rainfall regime west of 90° W. (Edmonton and St Louis) has a pronounced summer maximum associated with convectional showers. This is still evident as far east as *c*. 80° W., but less pronounced since winter frontal depressions are both more numerous and more active. The easternmost part of the region, through which most frontal depressions track throughout the year, has a fairly even distribution of rainfall, with a tendency towards a winter maximum (Halifax graph) as the Atlantic coast is approached.

With mean monthly temperatures below freezing point for up to $3\frac{1}{2}$ months on the Atlantic coast (Halifax) and 6 or more months in the Canadian prairies, much of the winter precipitation is in the form of snow. Snowfall ranges from an annual average 500 mm on the southern borders of the region and 750 mm in the Canadian prairies to almost 4000 mm in the uplands of the eastern part of the area.

The influence of the Great Lakes

Lakes are responsible for creating local climatic conditions within their vicinity, and this is particularly noticeable in the case of the Great Lakes which cover 243,000 sq km. The cooling of air in its passage over the lakes in summer tempers the heat where the winds are onshore; in winter onshore winds ameliorate the temperatures. Thus on one winter's day when south-west winds were blowing across Lake Michigan, Milwaukee recorded a temperature of $-18°$ c whilst on the opposite shore it was $-6°$ c.

By about mid-December the surface water temperature of the lakes has fallen to $4°$ c. Water has its maximum density at this temperature and overturning takes place with cooler water rising to the surface. Persistent ice formation then begins. Fig. 133(*a*) shows the extent of ice on 31 December 1966, about a fortnight after the overturning. The winter freeze of the Great Lakes reaches its maximum in February and considerable amounts of ice persist into mid-spring. Fig. 133(*b*) shows the ice concentration at the spring equinox of 1967 when only Lake Michigan and Lake Ontario were largely ice free.

The general amelioration of winter temperatures on the eastern shores of Lake Michigan reduces the winter killing of buds and allows the growing of a variety of fruits. Since the surface temperature of the lakes remains near 0° c throughout the period of ice-melt, blossoming of fruit trees is delayed until late spring. Then there is little chance of a killing frost. Conversely the onset of autumn frosts is later

133(*a*) Ice conditions on the Great Lakes, 31 December 1966
133(*b*) Ice conditions on the Great Lakes, 21 March 1967

134 Examples of E climate (northern hemisphere)

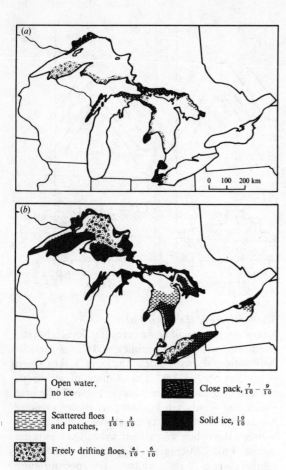

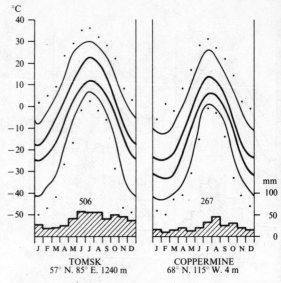

Open water, no ice

Close pack, $\frac{7}{10} - \frac{9}{10}$

Scattered floes and patches, $\frac{1}{10} - \frac{3}{10}$

Solid ice, $\frac{10}{10}$

Freely drifting floes, $\frac{4}{10} - \frac{6}{10}$

TOMSK
57° N. 85° E. 1240 m

COPPERMINE
68° N. 115° W. 4 m

along the lake shores than inland from them. The length of the frost-free season is also longer along the lake shores – in the Lake Peninsula of Ontario it is 154 days on the northern shores of Lakes Erie and Ontario but only 126 days in the interior of the peninsula.

Boreal climates (E)

The boreal climates (E) with less than four months over 10° C form broad belts across North America, Europe and Asia. In Canada its latitudinal extent is often 1600 km. In Finland and European Russia, where easier incursion of warmer maritime air pushes the D zone boundary northwards, its extension is half that of central Canada, but in Siberia, in the longitude of Lake Baikal, it widens again to 2000 km. Thus it covers an appreciable section of the northern continents. It is dominated by Arctic air masses particularly in winter. Because of the different distribution of land and sea it is entirely missing in the southern hemisphere.

In the U.S.S.R. the D/E boundary runs between

Omsk and Tomsk. Omsk, despite its severe winters and its low precipitation with a marked summer maximum, has four months over 10° C and so lies in the Dc zone. Tomsk (Fig. 134), with a similar regime but only three months over 10° C, just lies in the E zone. At Tomsk daily maxima range from −18° C in January to 23° C in July. In July 36° C has been recorded and so has 2° C, a range reflecting air masses from the hot interior and the arctic wastes respectively. Hard frost has occurred in every month except July and even July has had ground frost. The average daily minimum in January is −24° C, and −40° C can be expected in half the Januarys and Decembers. Precipitation totals 508 mm, with 65 mm falling in each of the three summer months (June, July and August); 51 mm falls in October and 48 mm in November. Much of this must be snow, giving a good cover for the rest of the winter. May is the spring month.

Archangel also has three months over 10° C and 506 mm of precipitation. It is somewhat less extreme than Tomsk.

In Canada the D/E boundary lies near the northern border of Alberta. Here Grand Prairie and Edmonton have five months reaching 10° C or above but Yellowknife in the North-west Territories only three months. Coppermine on the shore of the Arctic Ocean only reaches 10° C in July so is right on the northern border of the E zone. The average daily maximum varies from 14° C to −24° C and the minimum from 6° C to −43° C. An absolute maximum of 30° C has been recorded in summer but below freezing temperatures also occur in the summer months of June to August and the mean monthly temperature for July is 1° C. Precipitation is light with 267 mm and a late summer maximum.

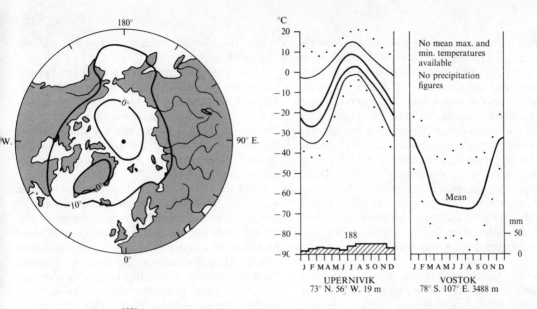

UPERNIVIK
73° N. 56° W. 19 m

VOSTOK
78° S. 107° E. 3488 m

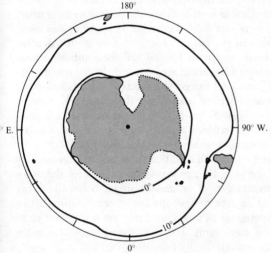

Polar climates (F)

Polar climates extend from the poles to regions where the average for the warmest month just fails to reach 10° C. It is important to remember that the early climatic classifications which used the tropics and polar circles to divide the earth into torrid, temperate and frigid zones resulted in very dissimilar sized areas.

Equator to tropics (torrid zone)	40%
Tropics to polar circle (temperate zone)	51%
Polar circle to pole (frigid zone)	9%

These percentages put the polar areas in perspective. The isotherm of 10° C for the warmest month is shown on Fig. 135. In the south the 10° C warmest-month isotherm rings the globe near 50° S. and encloses about 12% of the global surface. In the north the course of the isotherm is more complicated: in places it is well north of the arctic circle and only encloses 5–6% of the global surface – just about half of the F area in the southern hemisphere. Actually a better name might be high-latitude climate, which removes any confusion concerning the polar circles.

Poleward of the warmest-month isotherm of 10° C two climatic sub-types can be distinguished. Both are under the influence of arctic air masses throughout the year.

(*a*) Fi. Ice-cap climate occurs where the warmest month is 0° C or less. The largest area is Antarctica. The ice-cap interior of Greenland and the permanent pack-ice of the Arctic Ocean also belong here. Climatological records from the ice-cap regions are scarce and not often of long duration. For earlier years knowledge depends on exploratory expeditions and is sporadic. More recently long-term scientific bases have been established and useful information is accumulating. Vostok, the Russian base at 78° S., is an extreme example of this regime (Fig. 136). The average of the warmest

117

month is minus 30° c and the absolute maximum is minus 20° c. The peculiar curve of temperature is noteworthy. Once the sun has disappeared in April the outgoing radiation quickly brings the temperature to near its lowest value and several months of very similar conditions ensue. Conditions do fluctuate with the synoptic situation but average temperature varies little between May and September. Vostok's absolute minimum is −88° c in August, which has a daily mean of −68° c. At the south pole July, August and September have daily mean temperatures of −59° c and the lowest record is −80° c. With a longer observation period it seems likely that somewhere in the vicinity of the cold pole of Antarctica the −100° c will be registered one day.

On the Greenland ice-cap Eismitte (71° N.) recorded a mean daily minimum of −53° c for February 1931 and an absolute minimum of −65° c in March. The absolute maximum was −3° c, many degrees warmer than any temperature measured at Vostok.

(b) Ft. The tundra climate has a warmest month between 0° c and 10° c. Fig. 135 shows that this occurs on the northern fringes of Asia, Europe and North America, including the Canadian islands. In some places the strip is very narrow, as at Coppermine mentioned in the section on boreal climates. The coast lands of Greenland, except for small areas in the far north, are in the tundra. Peary Land (82° N.) has three months above 0° c, as has Alert on the northern tip of Ellesmere Land. Upernivik (Fig. 136) on the coast of Greenland at 73° N. has four months over 0° c but is less extreme than Coppermine which is just boreal (E). Because it receives maritime air masses at any time of the year it records temperatures not much below freezing on some occasions in the coldest months. February, for instance, has a monthly maximum of −2° c and has recorded +10° c. Conversely because it is very near the edge of the ice-cap the corresponding minima are −35° c and −42° c. Even in the warmest month the monthly minimum is −2° c and the absolute minimum −7° c.

In the southern hemisphere Grahamland peninsula is the only tundra area. On its northern tip, Deception Island (63° S.) has three months above 0° c with a mean of 1° c in the warmest month, while Stonington Island (68° S.) has a warmest month (January) mean of 0° c and so lies right on the tundra/ice-cap boundary. All the rest of Antarctica has an ice-cap regime.

Poleward of the isotherm of 10° c for the warmest month are large areas of ocean where the range of temperatures is small and it has been suggested that the area now labelled tundra should be subdivided by temperature range into a high-latitude marine climate where the range is less than 16° c, leaving the more extreme areas with a range above 16° c as the true regions of tundra climate. As its name implies, the high-latitude marine climate is mainly maritime in distribution – the environment of many whalers and fishermen – but it also includes some land areas. In the north the western Aleutian Islands, northern Iceland and the southern coastal areas of Greenland are included. In the south parts of Tierra del Fuego, the Falkland Islands and many isolated oceanic islands have this climate. Kerguelen Island (49° S. 70° E.) on the equatorward fringe of the high-latitude marine zone is one example. Its characteristics have been described as follows: 'It has a climate so cold that snow may fall on any day of the year, though it is warm enough for the mean temperature of no month to be below freezing; cold enough to have the lowest temperature below freezing on 248 days annually, yet warm enough to have more rainy days than snowy ones.' Compare this with the Scilly Isles at 50° N.

Though ice-cap, tundra and high-latitude marine areas have few inhabitants they are not unimportant. As sources of cold air and a sink of energy they have an essential role in maintaining the world balance of radiation.

Index